Tucholsky Wagner Zola Scott Sydow Freud Schlegel
Turgenev Fonatne Wallace
Twain Walther von der Vogelweide Fouqué Friedrich II. von Preußen
Weber Freiligrath Frey
Kant Ernst
Fechner Fichte Weiße Rose von Fallersleben Richthofen Frommel
Hölderlin
Fehrs Engels Fielding Eichendorff Tacitus Dumas
Faber Flaubert
Maximilian I. von Habsburg Eliasberg Ebner Eschenbach
Feuerbach Fock Zweig
Ewald Eliot Vergil
Goethe Elisabeth von Österreich London
Mendelssohn Balzac Shakespeare Dostojewski Ganghofer
Lichtenberg Rathenau Doyle Gjellerup
Trackl Stevenson Hambruch
Mommsen Tolstoi Lenz Droste-Hülshoff
Thoma Hanrieder
Dach von Arnim Hägele Hauff Humboldt
Verne Hauptmann
Karrillon Reuter Rousseau Hagen Gautier
Garschin Baudelaire
Damaschke Defoe Hebbel
Descartes Hegel Kussmaul Herder
Wolfram von Eschenbach Dickens Schopenhauer Rilke George
Bronner Darwin Melville Grimm Jerome Bebel
Campe Horváth Aristoteles Proust
Bismarck Vigny Voltaire Federer Herodot
Barlach
Gengenbach Heine
Storm Casanova Tersteegen Grillparzer Georgy
Lessing Gilm
Chamberlain Langbein Gryphius
Brentano Lafontaine
Strachwitz Claudius Schiller Kralik Iffland Sokrates
Bellamy Schilling
Katharina II. von Rußland Gerstäcker Raabe Gibbon Tschechow
Löns Hesse Hoffmann Gogol Wilde Vulpius
Luther Heym Hofmannsthal Gleim
Roth Klee Hölty Morgenstern Goedicke
Luxemburg Heyse Klopstock Kleist
Puschkin Homer
Machiavelli La Roche Horaz Mörike Musil
Navarra Aurel Musset Kierkegaard Kratt Kraus
Lamprecht Kind
Nestroy Marie de France Kirchhoff Hugo Moltke
Nietzsche Nansen Laotse Ipsen Liebknecht
Marx Ringelnatz
von Ossietzky Lassalle Gorki Klett Leibniz
May vom Stein Lawrence Irving
Petalozzi Knigge
Platon Kafka
Sachs Pückler Michelangelo Kock
Poe Liebermann
de Sade Praetorius Mistral Zetkin Korolenko

The publishing house tredition has created the series **TREDITION CLASSICS**. It contains classical literature works from over two thousand years. Most of these titles have been out of print and off the bookstore shelves for decades.

The book series is intended to preserve the cultural legacy and to promote the timeless works of classical literature. As a reader of a **TREDITION CLASSICS** book, the reader supports the mission to save many of the amazing works of world literature from oblivion.

The symbol of **TREDITION CLASSICS** is Johannes Gutenberg (1400 – 1468), the inventor of movable type printing.

With the series, tredition intends to make thousands of international literature classics available in printed format again – worldwide.

All books are available at book retailers worldwide in paperback and in hardcover. For more information please visit: www.tredition.com

tredition was established in 2006 by Sandra Latusseck and Soenke Schulz. Based in Hamburg, Germany, tredition offers publishing solutions to authors and publishing houses, combined with worldwide distribution of printed and digital book content. tredition is uniquely positioned to enable authors and publishing houses to create books on their own terms and without conventional manufacturing risks.

For more information please visit: www.tredition.com

Textiles For Commercial, Industrial, and Domestic Arts Schools; Also Adapted to Those Engaged in Wholesale and Retail Dry Goods, Wool, Cotton, and Dressmaker's Trades

William H. (William Henry) Dooley

Imprint

This book is part of the TREDITION CLASSICS series.

Author: William H. (William Henry) Dooley
Cover design: toepferschumann, Berlin (Germany)

Publisher: tredition GmbH, Hamburg (Germany)
ISBN: 978-3-8491-7365-4

www.tredition.com
www.tredition.de

COTTON PLANT

[Pg v]

PREFACE

The author established and since its inception has been in charge of the first industrial school for boys and girls in Massachusetts. At an early date he recognized the need of special text-books to meet the demand of young people who are attending vocational schools. There are plenty of books written on textiles for technical school students and advanced workers. But the author has failed to find a book explaining the manufacture and testing of textiles for commercial, industrial, domestic arts, and continuation schools, and for those who have just entered the textile or allied trades. This book is written to meet this educational need. Others may find the book of interest, particularly the chapters describing cotton, woolen, worsted, and silk fabrics.

The author is under obligations to Mr. Franklin W. Hobbs, treasurer of the Arlington Mills, for permission to use illustrations and information from literature published by the Arlington Mills; to Mr. S. H. Ditchett, editor of *Dry Goods Economist*, for permission to use information from his publication, "Dry Goods Encyclopedia"; [Pg vi] to the editor of the *Textile Mercury*; to Frank P. Bennett, of the *American Wool and Cotton Reporter*, for permission to use information from "Cotton Fabrics Glossary"; and to the instructors of the Lawrence Industrial School for valuable information. In addition, information has been obtained from the great body of textile literature, which the author desires to acknowledge.

[Pg vii]

CONTENTS

Carding, Combing. Worsted Tops—Gill Boxes. Different methods of Spinning—Bradford or English System, French System. Structure of Worsted Yarn. Uses of Worsted Yarn. Counts of Worsted Yarn

CHAPTER V
WOOLEN YARN

Operations in Producing Woolen Yarn—Washing, Carding, Spinning, Mule Spinning. Counts of Woolen Yarn. Uses of Woolen Yarn

CHAPTER VI
WEAVING

Preparatory to Weaving—Warp. Weaving—Weaving Processes, Classes of Weave—Plain or Homespun Weave, Twill, Satin Weaves, Figure Weaving (Jacquard apparatus), Double Cloth, Pile Weaving, Gauze Weaving, Lappet Weaving

CHAPTER VII
DYEING AND FINISHING

Dyeing. Wool Dyeing, Piece Dyed, Cross Dyed, Yarn Dyed. Style—Designing, Finishing, Perching, Burling, Mending, Fulling, Crabbing, Tentering, Napping, Pressing. Theories of Coloring in Textile Design. Various Methods of Employing Fancy Shades. Adulteration

CHAPTER VIII
WOOLEN AND WORSTED FABRICS

Albatross, Alpaca, Corded Alpaca, Angora, Astrakhan, Bandanna, Beaver (Fur Beaver), Bedford Cord, Beige, Bindings, Bombazine, Bottany, Boucle, Broadcloth, Bunting, Caniche, Cashmere, Cashmere Double, Cassimere, Castor, Challis, Cheviot (Diagonal or Chevron), Chinchilla, Chudah, Corduroy, Côte Cheval, Coupure, Covert, Delaine, Doeskin, Drap d'Été, Empress Cloth, Épingline, Etamine, Felt, Flannel, Dress Flannel, French Flannel, Shaker Flannel, Indigo Blue, Mackinaw, Navy Twilled Flannel, Silk Warp,

Baby Flannel. Florentine, Foule, Frieze, Gloria, Granada, Grena-
dine, Henrietta Cloth, Homespun, Hop Sacking, Jeans, Kersey,
Kerseymere, Linsey Woolsey, Melrose, Melton, Meltonette, Merino,
Mohair Brilliantine, Montagnac, Orleans, Panama Cloth, Prunella,
Sacking, Sanglier, Sebastopol, Serge, Shoddy, Sicilian, Sultane,
Tamise, Tartans, Thibet, Tricot, Tweed, Veiling, Venetian, Vigogne
(Vicuña), Vigoureux, Voiles, Whipcord, Worsted Diagonals, Zeph-
yr, Zibeline

CHAPTER IX

COTTON

Rough Peruvian, East Indian, Egyptian, Sea Island. American
Crop—Planting, Picking, Ginning—Roller Gins, Saw Gins. Cotton
Gin. Information on the Leading Growths of Cotton. Grades—Full
Grades, Half Grades, Quarter Grades. Varieties—Sea Island (select-
ed), Sea Island (ordinary), Florida Sea Island, Georgia, Egyptian,
Peeler, Orleans or Gulf Upland, Texas

CHAPTER X

MANUFACTURE OF COTTON YARN

Picker Room, Carding Machine, Combing, Drawing. Flyer
Frames—Intermediate Frame, Roving Frame, Fine or Jack Frame
Spinning—Mule Spinning, Ring Spinning

CHAPTER XI

THREAD AND COTTON FINISHING

Manufacturing Processes. Thread Numbers. Sizing. Cotton Finishing—Bleaching, Starching, Calendering, Mercerizing. Characteristics of fine Cotton Cloth

Acid Test. Cotton Distinguished from Linen, Silk from Wool, Artificial Silk from Silk. Test for Shoddy. Determination of Dressing. Test for Permanence of Dyes

History of Textiles. History of the Organization of Textile Industries. History of Manufacturing. History of Lace

EXPERIMENTS

Experiments 1 to 62

Sources of Supply

[Pg 1]

TEXTILES

CHAPTER I

FIBERS

All the materials used in the manufacture of clothing are called *textiles* and are made of either long or short fibers. These fibers can be made into a continuous thread. When two different sets of threads are interlaced, the resulting product is called cloth.

The value of any fiber for textile purposes depends entirely upon the possession of such qualities as firmness, length, curl, softness, elasticity, etc., which adapt it for spinning. The number of fibers that possess these qualities is small, and may be classified as follows:

Animal Fibers: Wool, Silk, Mohair.

Vegetable Fibers: Cotton, Flax, Jute, Hemp, etc.

Mineral Fibers: Asbestos, Tinsel, and other metallic fibers.

Remanufactured Material: Noils, Mungo, Shoddy, Extract, and Flocks.

Artificial Fibers: Spun Glass, Artificial Silk, and Slag Wool.

The Structure of Wool. A large part of the people of the world have always used wool for their clothing. Wool is the soft, curly covering which forms the fleecy [Pg 2] coat of the sheep and similar animals, such as the goat and alpaca. Wool fiber when viewed under the microscope is seen to consist roughly of three parts:

1st. Epidermis, or outer surface, which is a series of scales lying one upon the other.

2d. Cortex, or intermediate substance, consisting of angular, elongated cells, which give strength to the wool.

3d. Medulla, or pith of the fiber.

WOOL FIBER

Highly magnified

Difference between Wool and Hair. Not all animal fibers are alike. They vary in fineness, softness, length, and strength, from the finest Merino wool to the rigid bristles of the wild boar. At just what point it can be said that the animal fiber ceases to be wool and becomes hair, is difficult to determine, because there is a gradual and imperceptible gradation from wool to hair. [1] The distinction between wool and hair lies chiefly in the great fineness, softness, and wavy delicacy of the woolen fiber, combined with its highly serrated surface — upon which the luster of the wool depends.

Characteristics of Wool. The chief characteristic of wool is its felting or shrinking power. This felting property from which wool derives much of its value, and which is its special distinction from hair, depends in part upon the kinks in the fiber, but mainly upon the [Pg 3] scales with which the fiber is covered. These scales or points are exceedingly minute, ranging from about 1,100 to the inch to nearly 3,000. The stem of the fiber itself is extremely slender, being less than one thousandth of an inch in diameter. In good felting wools the scales are more perfect and numerous, while inferior wools generally possess fewer serrations, and are less perfect in structure.

In the process of felting the fibers become entangled with one another, and the little projecting scales hook into one another and hold the fibers closely interlocked. The deeper these scales fit into one another the closer becomes the structure of the thread.

Classification of Wool. The various kinds of wool used in commerce are named either from the breed of the sheep or from the country or locality in which the sheep are reared. Thus we get Merino wool from Merino sheep, while English, American, and Australian wools are named from the respective countries. As the result of cross breeding of different sheep in different parts of the world, under different climatic conditions, physical surroundings, and soil, there exist a great many varieties of wool. The wool of commerce is divided into three great classes: (1) Short wool or clothing wool (also called carding wool), seldom exceeding a length of two to four inches; (2) long wool or combing wool, varying from four to ten inches; (3) carpet and knitting wools, which are long, strong, and very coarse.

The distinction between clothing or carding wools on the one hand, and combing wools on the other, is an [Pg 4] old one. Combing wools are so called because they are prepared for spinning [2] into yarn by the process of "combing"—that is, the fibers are made to lie parallel with one another preparatory to being spun into thread. Carding wools—made to cross and interlace and interlock with one another—are shorter than combing wools, and in addition they possess to a much greater degree the power of felting—that is, of matting together in a close compact mass. Combing wools, on the other hand, are not only longer than the carding wools, but they are also harder, more wiry, and less inclined to be spiral or kinky. It must be understood, however, that under the present methods of manufacture, short wools may be combed and spun by the French method of spinning just as the long wools are combed and spun by the Bradford or English system.

Carpet and knitting wools are the cheapest, coarsest, and harshest sorts of wools. They come principally from Russia, Turkey, China, Greece, Peru, Chili, etc., and from the mountain districts of England and Scotland. Carpet wools approach more nearly to hair than other wools. The only staple of this class produced in the United States is grown on the original Mexican sheep of the great Southwest. Few of these Mexican sheep are left, for they have been improved by cross breeding, but they constitute the foundation stock of most of our Western flocks, which now produce superior clothing and combing wool.

[Pg 5] **Sheep Shearing.** In order to get an idea of the importance of the sheep industry in the United States, one must take a glance at its condition in the big states of the West. Wyoming has more than 4,600,000 sheep within its borders. Montana, which held the record until 1909, has 4,500,000 sheep. Then comes Idaho with 2,500,000, Oregon with 2,000,000, and so on down the list until the nation's total reaches 40,000,000 sheep, four-fifths of which are west of the Missouri river.

SHEEP SHEARING

[Pg 6] To harvest the wool from such an enormous number of backs is a task that calls for expert shearers, men who can handle the big shears of the machine clippers with a skill that comes from long practise. The shearing must be done at the right time of the year. If the wool is clipped too early, the sheep suffer from the cold; if the shearing comes too late, the sheep suffer from intense heat, and in either case are bound to lose weight and value.

To meet the exacting conditions a class of men has risen expert in the sheep-shearing business. These shearers begin work in southern and middle California, Utah, etc. Another month finds them busy in the great sheep states of Wyoming, Montana, Idaho, and Oregon, where they find steady employment until July, when they go to the ranges of Canada. In this way the shearers keep busy nearly all the year, and at high wages.

The Mexicans are particularly expert with the hand shears, though this form of clipping is being done away with, owing to the installation of power plants for machine shearing. These plants are installed at various points on the great sheep ranges. Long sheds are erected and shafting extends down both sides of the shearing place. Twenty or more shearers will be lined up in one of these sheds, each man operating a clipping machine connected with the shafting. The sheep are brought in from the range in bands of 2,500 or more, and are put in the corrals adjoining the shearing sheds. Then they are driven down chutes to the shearers.

[Pg 7] A shearer reaches into a small corral behind him and pulls out a sheep. With a dexterous fling the animal is put in a sitting posture between the shearer's knees, and then the steel clippers begin clipping off the wool. The machine-shearing saves much wool, as it gets closer to the skin of the sheep and shears more evenly. In fact, some sheep owners say that the increased weight of their fleeces at each shearing is enough to pay the extra expense of running a power plant.

As fast as the sheep are turned out by the shearers they are run along a narrow chute and each one is branded. The branding mark is usually a letter painted on the back of the sheep so that it can be plainly seen when they are coming through a chute. The mark remains on the fleece and is always easily distinguished.

Fleece. There is a great variation in the weight of fleeces. Some sheep, such as those on the best ranges in Oregon, Montana, and Wyoming, will average an eight-pound fleece full of natural oil, while sheep from the more sterile alkaline ranges of New Mexico will not average much more than five pounds of wool.

The shearing season on the plains is much like the threshing season in agricultural communities. With a crew of first-class shearers working in a shearing shed, it is not long until the floor is a sea of wool. Boys are kept busy picking up the fleeces, tying them into compact bundles, and throwing them to the men who have been assigned to the task of filling the wool sacks. These sacks, which hold about 400 pounds, are [Pg 8] suspended from a wooden framework, and as fast as the fleeces are thrown in, they are tramped down until the sacks will not hold a pound more. Most of the sacks are shipped to warehouses in such wool centers as Casper, Wyoming, or Billings, Montana, the latter place being the greatest wool shipping center in the world. Here they are sold to Eastern buyers, who examine the clips at their leisure and make their bids.

Value of Wool Business. Some idea of the fortunes at stake in the wool business can be gathered from the fact that the total wool product of the country in 1909 was valued at $78,263,165. It is expected that the returns from the wool clip in a fairly good year will pay all a sheepman's running expenses, such as hire of herders, cost of shearing, etc. He then has the sale of his lambs as clear profit. Enormous fortunes are being made in the sheep business in the west, owing to the high price of wool and mutton.

Saxony and Silesian Wool. Among wools of all classes the Saxony and Silesian take the first place, and for general good qualities, fineness, and regularity of fiber, they are unequalled. The fiber is short in staple, possesses good felting properties, and is strong and elastic. This wool is used chiefly in the manufacture of cloths where much milling [3] is required, such as superfines and dress-faced fabrics.

Australian Wools. Australia furnishes wools of a superior character, and some of the choicest clips rival [Pg 9] the Saxony and Silesian wools. They are used both for worsted [4] and woolen yarns. They are generally strong and of an elastic character, possess numerous serrations, and are of good color, with good felting properties. The principal Australian wools are Port Philip, Sydney, and Adelaide wools. These are the best brands imported from that country.

Port Philip Wool. Port Philip wool is suitable for either worsted or woolen yarns. The fiber is not quite as fine as Saxony, but it makes a good thread, is fairly sound in staple, and is of good length and color. It is very wavy and serrated. The longest and best of this wool is used for the very finest worsted yarns, and will spin up to 130's counts. [5] The sheep are descendants of the original Spanish Merino. Cross bred Port Philip wool is from the same Merino sheep crossed with Leicesters, which yield a medium quality fleece of sound fiber and good quality for spinning counts from 40's to 56's. The yarn has a bright, clear appearance.

Sydney Wools. Sydney wools are moderately fine in fiber and of medium length. They are rather deficient in strength, uneven in color, and often contain yellow locks which make them undesirable when required for dyeing light shades. They are used for nearly the same purpose as Port Philip wools, but do not spin quite as far in worsted yarns, nor are they equal in milling qualities.

[Pg 10] **Adelaide Wool.** Adelaide wool has a reputation for sound Merinos, the average quality being a little lower than for the Port Philip and Sydney wools. Its fiber is moderately fine, but not of uniform length; its color is not so good, and it contains a large amount of yolk. [6] Adelaide wool is used for worsted dress goods, weft (filling) [7] yarn up to 60's, and certain worsted warps. [7] It is used for medium fancy woolens.

Van Wool from Tasmania. The climate of this island is well suited to the growing of wool, and produces excellent qualities, fine in fiber, of good length, and strong in the staple, which will spin as high counts as 70's and 80's worsted. This wool is useful for mixing with other good wools. Its color is very white, which makes it a useful wool for dyeing light shades. Its milling properties are good, and the shorter sorts are suitable for woolens.

New Zealand Wools are very supple, which make them valuable to the spinner. These wools are suitable for almost all classes of Merino and crossbred yarns. They are of good length, sound staple, have good felting properties, and are of good color. They are useful for blending with mungo and shoddy, to give to these remanufactured materials that springy, bulky character which they lack.

Cape Wools. Cape Colony and Natal produce merino wool that is somewhat short in staple, rather tender, and less wavy than some other wools. The [Pg 11] sheep are not so well cared for, and are fed on the leaves of a small shrub. The absence of grass leaves the ground very sandy, and this makes the fleece heavy and dirty. Its color is fair, but it lacks elasticity. It is used chiefly to cheapen blends [8] of 60's top. [9] The short wool is combed for thick counts for weft and hosiery, and is also used for shawls and cloths where felting is not an essential feature.

MERINO SHEEP

Wools from South America. These wools are of the same standard of excellence as the Australian wools, but they are generally deficient in strength and elasticity. Buenos Ayres and Montevideo wools are fairly fine in fiber, but lack strength and elasticity, and are deficient in milling properties; they are also [Pg 12] burry. The climate suits the sheep well, and the feed is good, but the careless methods of classing and packing have earned for these wools a poor reputation that is well deserved.

The best 60's wool is combed in oil, but a large portion of the shorter is combed and used in thick counts,—20's to 36's worsted for the hosiery trade.

Russian Wool. The staple of this is generally strong, and the fibers are of a medium thickness; the color is milky white. It is useful to blend with Australian or other good wools. It produces a good yarn, and is very often used in the fancy woolen trade and in fabrics that require to be finished in the natural color.

Great Britain Wools. These may be divided into three groups: (1) long wools, of which the Lincoln and Leicester are typical examples; (2) short wools, which include Southdown, Shropshire, Suffolk, and others; and (3) wool from the mountain or hilly breeds of sheep, such as the Cheviot, Scotch Blackface, Shetland, Irish, and Welsh.

Lincoln Wool is a typical wool obtained from the long wool sheep, and noted for its long, lustrous fiber, which is silky and strong. The staple varies from ten to eighteen inches in length, and the average fleece will yield from ten to fourteen pounds in weight.

Leicester Wool has a somewhat finer fiber than Lincoln. It is a valuable wool, of good color, uniform and sound in staple, curly, with good, bright luster and no dark hairs. While luster wools are grown extensively in England, they also grow in Indiana and [Pg 13] Kentucky, and are commonly known in the trade as braid wool.

Southdown is one of the most valuable of short staple wools. It possesses a fine hair, is close and wavy, and fairly sound in staple, but rather deficient in milling qualities. The shorter varieties are carded and made into flannels and other light fabrics, while the longer qualities are used in the production of worsted goods. The weight of a Southdown fleece averages from four to five pounds.

WOOL MARKET AT BUENOS AYRES

Shropshiredown wool is of good quality, with strong, fine, lustrous fiber, of good length. It resembles Southdown, but is not as

24

lustrous as mohair, the natural colors being either white, black, brown, or fawn. It is used chiefly in the manufacture of dress goods.

Cashmere Wool is the fine, woolly, extremely soft, white or gray fur of the Cashmere goat which is bred in Thibet. There are two kinds of fiber obtained: one, [Pg 14] which is really the outer covering, consists of long tufts of hair; underneath this is the Cashmere wool of commerce, a soft, downy wool of a brownish-gray tint, with a fine, silky fiber. It is used for making the costly oriental (Indian) shawls and the finest wraps.

The Norfolkdown and Suffolkdown Wools are fairly fine in fiber and soft, but slightly deficient in strength and elasticity.

Cheviot Wool may be taken as representative of the hilly breeds of sheep. It is an average wool, with staple of medium length, soft, and with strong and regular fiber; it is of a good, bright color, and possesses desirable milling properties, being used for both woolen and worsted, but chiefly in the fancy woolen trade. The average weight of the fleece is about 4½ pounds. The black-faced or Highland breed yields a medium wool, coarser and more shaggy than the Cheviot, and varying much in quality. It is almost all used in the production of rugs, carpets, and blankets.

Welsh Wools lack waviness and fineness of fiber. They are chiefly used for flannels.

Shetland Wools are similar in character to Welsh wools, but slightly finer in fiber and softer. They are used in the manufacture of knitted goods, such as shawls and wraps. They lack felting properties.

Irish Wools possess a strong, thick hair of moderate length and fine color. They are similar in many respects to the Welsh wools, and are often classed with them. They are used in the production of low and [Pg 15] medium tweeds—fancy woolen cloths not requiring small yarns or milling qualities.

Mohair is a lustrous wool obtained from the Angora goat, which derives its name from the district of Asia Minor from which it comes. These animals have also been successfully bred in Spain and France. The hair is pure white, fine, wavy, and of good length, and possesses a high luster. It is used in making plushes, velvets, astrakhans, and curled fabrics, also half silk goods and fine wraps.

Alpaca Wool is the fleece of the Peruvian sheep, which is a species of llama. The staple is of good length and soft, but is not quite as lustrous as mohair, the natural colors being either white, black, brown, or fawn. It is used chiefly in the manufacture of dress goods.

How Wool is Marketed. The bulk of the wool of commerce comes into the market in the form of fleece wool, the product of a single year's growth, and cut from the body of the animal usually in April or May. The first and finest clip, called lamb's wool, may be taken from the young sheep at the age of eight to twelve months. All subsequently cut fleeces are known as wether wool and possess relatively somewhat less value than the first clip.

FOOTNOTES:

[1] Hair is straight and glossy, stronger and smoother than wool, and grows sometimes as long as twenty inches.

[2] Spinning is a process by which long or short fibers are twisted into a continuous thread.

[3] A process of finishing cloth by condensing the fibers so as to make the cloth stronger and firmer.

[4] See footnote, page 39.

[5] The size of yarn is technically called the "counts" and is based on the number of 560 yard lengths required to weigh one pound. In this case 130's count = 130 × 560, or 72,800 yards of yarn to a pound.

[6] An encrusting compound of dirt and grease formed on the fleece.

[7] See page 54.

[8] Mixtures.

[9] After wool fibers are combed they are called top.

[Pg 16]

CHAPTER II

WOOL SORTING

Fleece wool as it comes to the mill is rolled up in bundles and must be sorted. This process consists in sorting and classifying the fibers of the fleece. Not only do the various species of sheep furnish widely different qualities of wool, but different qualities are obtained from the same animal, according to the part of the body from which the wool is taken. This variation in some instances is very marked, and sometimes is greater than that which separates the wools of the different breeds of sheep. Hence the sorting and classing of wool become necessary for the production of good, sound yarn of even quality.

An attempt to utilize the fleece as a whole would result in the spinning of uneven, faulty, and unsatisfactory yarns. As many as twelve or fourteen sorts may be obtained from one fleece (by very fine sorting), but generally not more than five or seven are made. The following table shows the relative qualities of wools from the various parts of a Merino sheep:

1 and 2. *Head* (*top and sides*): The wools grown on these parts are remarkable for length of staple, softness, and uniformity of character. They are usually the choicest wools in the fleece.

[Pg 17] 3. *Upper part of the back:* This also is a wool of good, sound quality, resembling in staples Nos. 1 and 2, but not as soft or as fine of fiber.

4. *Loin and back:* The staple here is comparatively short, not as fine, but generally of unvarying character, sometimes rather tender.

SORTING ROOM

5. *Upper parts of legs:* This wool is medium in length but coarse of fiber, and has a tendency to hang in loose, open locks. It is generally sound, but likely to contain vegetable matter.

6. *Upper portion of the neck:* The staple clipped from this part of the neck is of an inferior quality, frequently faulty and irregular in growth, and contains twigs, thorns, etc.

[Pg 18] 7. *Central part of the neck:* This wool is similar to No. 6 but rather tender in staple.

8. *Belly:* This wool is from under the sheep, between the fore and hind legs. It is short and dirty, poor in quality, and generally tender.

9. *Root of tail:* In this wool the fibers are coarse, short, and glossy.

10. *Lower parts of the legs:* This wool is generally dirty and greasy, the staple having no wave and lacking fineness. It is generally burry and contains much vegetable matter.

11. *Front of Head;* 12. *Throat;* 13. *Chest:* The wools from these parts are sometimes classed together, all having the same characteristics. The fiber is stiff, straight, coarse, and covered with fodder.

14. *Shins:* This is another short, thick, straight wool of glossy fiber, commonly known as shanks.

Classing. Classing is a grading of the fleeces, and is usually, but not always, a process preliminary to sorting. It is an important part of sorting, and when well done greatly facilitates the making of good, uniform matchings.

Grades of Wool. In the grading of wool no set standard of quality exists. The same classification may be applied in different years, or in different localities, to qualities of wool showing much variation, the best grade obtainable usually setting the standard for the lower grades. The highest quality of wool in the United States is found on full-blooded Merino sheep.

Merino Wool. The Merino sheep was bred for wool [Pg 19] and not mutton. The fleece of this breed is fine, strong, elastic, and of good color; it also possesses a high felting power. Though naturally short, it is now grown to good length and the fleece is dense. The Merino sheep is a native of Spain, and Spain was for a long period the chief country of its production. It was also in past centuries extensively bred in England and English wool owes much to the Merino for the improvement it has effected in the fleeces of other breeds of English sheep. It was also introduced into Saxony and was highly bred there, and Saxony soon came to surpass Spanish wool in fineness, softness, and felting properties. The Merino was introduced into the United States at the beginning of the nineteenth century. By 1810, 5,000 Merino sheep had been imported and these 5,000 sheep formed the basis of most of the fine wool-producing flocks of our country to-day.

The terms half blood, three-eighths blood, and quarter blood refer to the full-blooded Merino standard. As the scale descends the wool becomes coarser, the wool of a quarter blood usually being a comparatively coarse fiber. The general classifications of fine, medium, coarse, and low, refer to the relative fineness of Merino combing wools. These distinctions naturally overlap according to the opinion of the parties in transactions. Picklock XXX and XX represent the highest grades of clothing wool, the grade next lower being X, and then Nos. 1 and 2. These again are used in connection with the locality from which the wool is grown, as Ohio XX, Michigan X, New York No. 1, etc.

[Pg 20] **Difference Between Lamb's and Sheep's Wool.** One of the first points to be understood in wool sorting is the difference between the wool of lambs and one-year-old sheep, and that of sheep two or more years old. Lamb's wool is naturally pointed at the end, because it has never been clipped. It is termed hog or hoggett wool, and is more valuable when longer, of about fourteen months' growth. It is finer in quality and possesses more waviness, which is a help in the process of spinning.

The wool of sheep two or more years old is known as wether. The ends of the fiber from such sheep are thick and blunted, on account of having been previously cut. It is necessary to be able to tell at once a hog fleece from a wether, and this can be done in two ways: by examining the ends of the fiber to see if they are pointed; or by pulling a staple out of the fleece. If it is wether, the staple will come out clean, without interfering to any extent with those around it; but if hog, some of the fibers will adhere to the one that is being pulled. Hog wool is generally more full of dirt, moss, straw, and other vegetable matter.

Dead Wool is wool obtained from the pelts of sheep that have died.

Pulled Wool. Pulled wool is wool from the pelts [10] as they come from the slaughter-houses of large packing plants. These pelts are thrown into vats of water and left to soak for twenty-four hours to loosen the dirt which has become matted into the wool. From these [Pg 21] vats the pelts are taken to scrubbing machines from which the wool issues perfectly clean and white. The pelts are next freed from any particles of flesh or fat which may have adhered to them, and are then taken to the "painting" room. Here they are laid flesh side up and carefully painted with a preparation for loosening the roots of the wool. This preparation is allowed to remain on the pelts for twenty-four hours, when it is cleaned off and the pelts taken to the "pulling" room. Each wool puller stands before a small wooden framework over which the pelt is thrown, and the wool, being all thoroughly loosened by the "paint" preparation, is easily and quickly pulled out by the handful. As it is pulled it is thrown into barrels conveniently arranged according to grade and length.

When a barrel is filled, it is transferred to a large room across which are several rows of wire netting, raised about three feet from the floor. Each sheet of netting is about six feet wide. Here the wool is piled on the netting to a depth of several inches and hot air is forced underneath it by means of a blower. Meanwhile it is worked over by men with rakes, and soon dries. When thoroughly dry, it is raked up and taken to the storeroom, where it is dumped into bins. Here it usually remains open for inspection and sampling till it is sold, when it is bagged. The bags of wool ultimately find their way to the woolen mill or sampling house. Sometimes the fleece will retain its fleece form, but usually it breaks up. The wool contains lime and has to be specially treated by a scouring process to prevent lime [Pg 22] from absorbing the cleansing substances used for scouring the wool.

Delaine Wool is a variety of fine, long combing wool.

Cotty Wool, or cotts, is wool from sheep that have been exposed to severe weather and lack of nourishment, and for these reasons have failed to throw off the yolk necessary to feed the wool. As a result it becomes matted or felted together, and is hard and brittle and almost worthless.

Wool Sorter. The sorter begins by placing the fleece upon his board or table, always arranging it so that he faces the north, as this gives the most constant light and no glare of the sun. The fleece thus spread out shows a definite dividing line through the center. The sorter parts the two halves and proceeds to analyze their different qualities. The number of sorts is determined by the requirements of the manufacturer who, in purchasing his wool, buys those grades that will produce the greatest bulk of the qualities for present use, and that leave in stock the smallest number of sorts and least weight for which he has no immediate use. The sorter then removes all extraneous matter adhering to the fleece, such as straw, twigs, and seeds, and cuts off the hard lumps of earth, tar, or paint, which, if not removed at this time, will dissolve in the scouring process and stain the wool. With these preliminaries finished, he proceeds to cast out the locks, according to quality, into baskets or skeps provided for that purpose. After skirting or taking off the outside edges of the fleece, usually known as brokes, and [Pg 23] the legs and tail, known as breech, he separates the other portions from the better qualities.

SORTED WOOL IN PILES READY TO BE TRANSPORTED TO THE DEGREASING PLANT

Picklock, prime, choice, super, head, downrights, seconds, breech, etc., are some of the terms used. Picklock comprises the choicest qualities; prime is similar to picklock, but slightly inferior; choice is true staple, but not as fine in fiber; super is similar to choice, but as a rule not as valuable; head includes the inferior sorts from this part of the sheep; downrights come from the lower parts of the sides; seconds consist of the best wool clipped from the throat and breast; breech, the short, coarse fibers obtained from the skirting and edgings of the fleece.

In the worsted trade different names are used. The [Pg 24] terms generally adopted are: blue, from the neck; fine, from the shoulders; neat, from the middle of the sides and back; brown-drawings, from the haunches; breech, from the tail and hind legs; cowtail, when the breech is very strong; brokes, from the lower part of the front legs and belly, which are classed as super, middle, and common.

Fine, short wools are sorted according to the number of counts of yarn they are expected to spin, as 48's, 60's, 70's, and so on. Thus we see there are different methods of indicating qualities in different districts, and also of indicating differences of qualities between the woolen and worsted branches of the trade.

It may be noted that the quality of the wool varies in the same way as the quality of the flesh. The shoulder is finest in grain and most delicate, so the wool is finer in fiber. There is more wear and tear for the sheep at its haunches than at its shoulders, hence the wool is longer and stronger; about the neck the wool is short, to prevent the sheep from being weighted down while eating, etc.; the wool on the back becomes rough and thin, being most exposed to the rain. From the foregoing it will be readily seen that there is necessity for careful sorting, in order to insure obtaining an even running yarn, and subsequently a uniform quality of fabrics.

Wool Washing. Fleece wool as it comes into the market is either in the "grease," that is, unwashed and with all the dirt which gathers on the surface of the greasy wool; or it is received as washed wool, the [Pg 25] washing being done as a preliminary step to the sheep shearing. Wool, unlike cotton, cannot be worked into yarn without being thoroughly cleansed of its impurities. These impurities consist of greasy and sweaty secretions, of the nature of a lubricant to the fiber. Combined with dirt, sand, etc., which adhere to the wool, these secretions form an encrusting compound, known as yolk, which acts as a natural preservative to the wool, keeping it soft and supple. This compound, with other extraneous matter, must be removed before the wool is in a workable condition. The amount of yolk varies, the greatest amount being found in fine, short wools from the warm climates. In long-staple wool the amount of yolk is comparatively small.

WASHING ROOM

[Pg 26] Various methods of removing these impurities have been tried; one is the use of absorbent substances, such as fossil meal, alumina, etc., to withdraw the greasy matter, so that the remaining impurities can be easily removed by washing. In other methods, naphtha or similar solvent liquids are used to dissolve the wool fats. This is followed by washing in tepid water to dissolve the potash salts, leaving the dirt to fall away when the other substances are no longer present. To work this method with safety requires a costly and intricate plant with skilled supervision. The method which is practically in universal use is washing the wool in alkaline solutions, properties of which combine with and reduce the impurities to a lathery emulsion which is easily washed off from the wool.

Great stress is laid upon the necessity of care in the washing process, as the luster may be destroyed and a brownish-yellow tint given to the wool, the spinning properties very seriously injured, the softness destroyed, or the fiber dissolved. Some wools are easy to wash, requiring little soap and a reasonable temperature; other wools are cleansed with great difficulty. A note, therefore, should be made of any particular brand or class of wool requiring special attention, to serve as a guide in the treatment of future lots. The danger lies in using unsuitable agents,—hard water, excessive temperatures, strong reagents, etc.

Caustic alkalies have a most destructive effect on wool as they eat into it and destroy its vitality. Carbonate alkalies are less severe. Whatever cleansing [Pg 27] substances are used, it is essential that they should be free from anything that is likely to injure the wool—that they remove the impurities and still preserve all the qualities in the wool. If the washing is properly performed the alkaline portion of the yolk is removed, leaving only the colorless animal oil in the fiber. If the work is not thoroughly done the wool passes as "unmerchantable washed." "Tub washed" is the term applied to fleeces which are broken up and washed more or less by hand. Scoured wool is tub washed with warm water and soap, and then thoroughly rinsed in cold water until nothing remains but the clean fiber.

DEGREASING PLANT—REMOVING GREASE FROM WOOL

An improved method of washing wool by hand is to have a series of tanks with pressing rollers attached to [Pg 28] each tank: the wool is agitated by means of forks, and then passed to the pressing rollers and into each tank in succession. The tanks are usually five in number, and so arranged that the liquor can be run from the upper to the lower tank. Upon leaving the pressing rollers the excess of water is driven off in a hydro extractor [11] and the wool is beaten into a light, fluffy condition by means of a wooden fan or beater.

Wool Drying. The process of drying wool is not intended to be carried to such an extent that the wool will be in an absolutely dry state, for in such a condition it would be lusterless, brittle, and discolored. It is the nature of wool to retain a certain amount of moisture since it is hygroscopic, and to remove it entirely would result in partial disintegration of the fibers. Buyers and sellers have a recognized standard of moisture, 16 per cent. If, on the other hand, it is left too wet, the fibers will not stand the pulling strain in the succeeding operations, and if not broken, they are so unduly stretched that they have lost their elasticity.

The theory which underlies the drying process is that dry air is capable of absorbing moisture, hence by circulating currents of dry air in and around wet wool, the absorbing power of the air draws off the moisture. For continuous drying free circulation is a necessity, as otherwise the air would soon become saturated and incapable of taking up more moisture. Warming the air increases its capacity to absorb moisture; thus a [Pg 29] higher temperature is capable of drying the wool much quicker than the same volume of air would at a low temperature. A free circulation of air at 75 to 100 degrees F., evenly distributed, and with ample provision for the escape of the saturated air, is essential for good work.

Oiling. After being scoured wool generally has to be oiled before it is ready for the processes of spinning, blending, etc. As delivered from the drying apparatus, the wool is bright and clean, but somewhat harsh and wiry to the touch, owing to the removal of the yolk which is its natural lubricant. To render it soft and elastic, and to improve its spinning qualities, the fiber is sprinkled with lard oil or olive oil. As the oil is a costly item, it is of consequence that it be equally distributed and used economically. To attain this end various forms of oiling apparatus have been invented, which sprinkle the oil in a fine spray over the wool, which is carried under the sprinkler by an endless cloth.

Burring and Carbonizing. After wool has been washed and scoured it frequently happens that it cannot be advanced to the succeeding operations of manufacture because it is mixed with burs, seeds, leaves, slivers, etc., which are picked up by the sheep in the pasture. These vegetable impurities injure the spinning qualities of the stock, for if a bur or other foreign substance becomes fastened in the strand of yarn while it is being spun, it either causes the thread to break or renders it bunchy and uneven. For removing burs, etc., from the wool two methods are pursued: the [Pg 30] one purely mechanical, the other chemical, and known respectively as burring and carbonizing.

Bur Picker. For the mechanical removing of burs a machine called the bur picker is employed. In this machine the wool is first spread out into a thin lap or sheet; then light wooden blades, rotating rapidly, beat upon every part of the sheet and break the burs into pieces. The pieces fall down into the dust box or upon a grating beneath the machine, and are ejected together with a good deal of the wool adhering to them. Often the machine fails to beat out fine pieces and these are scattered through the stock.

Process of Carbonizing. For the complete removal of all foreign vegetable substances from wool the most effective process is carbonizing, in which the burs, etc., are burned out by means of acid and a high degree of heat. The method of procedure is as follows: The wool to be treated is immersed in a solution of sulphuric or hydrochloric acid for about twelve hours, the acid bath being placed in cement cisterns or in large lead-lined tubs and not made strong enough to injure the fiber of the wool. During the immersion the stock is frequently stirred. Next, the wool is dried and then placed in an enclosed chamber and subjected to a high temperature (75 degrees C.). The result of this process is that all the vegetable matter contained in the wool is "carbonized" or burned to a crisp, and on being slightly beaten or shaken readily turns to dust. This dust is removed from the wool by various simple processes. The carbonizing process was first introduced [Pg 31] in 1875, though it made but slight headway against the old burring method until after 1880.

Blending. Pure wool of but one quality is not often used in the production of woven fabrics, so, before the raw material is ready for spinning into yarn, or for other processes by which it is worked into useful forms, it is blended. Wools are blended for many reasons (among which cheapness figures prominently), the added materials consisting usually of shoddy, mungo, or extract fibers. Ordinarily, however, blending has for its object the securing of a desired quality or weight of cloth. The question of color, as well as quality, also determines blending operations, natural colored wools being frequently intermixed to obtain particular shades for dress goods, tweeds, knitting yarns, etc. Stock dyed wools are also blended for the production of mixed colors, as browns, grays, Oxfords, etc. There is practically no limit to the variety of shades and tints obtainable by mixing two or more colors of wool together. The various quantities of wool to be blended are spread out in due proportion in the form of thin layers, one on top of the other, and then passed through a machine called the teaser. The teaser consists of a combination of large and small rollers, thickly studded with small pins, which open the wool, pull it apart, and thoroughly intermix it. A blast of air constantly plays upon the wool in the teaser and aids the spikes and pins in opening out the fibers. The material is subjected to this operation several times and is finally delivered in a soft, fleecy condition, ready to be spun into yarn.

FOOTNOTES:

[10] Skins.

[11] A wire cage enclosed in a metallic shell which revolves at a high speed causing sixty or seventy per cent of the moisture to be removed.

[Pg 32]

CHAPTER III

WOOL SUBSTITUTES AND WASTE PRODUCTS

Remanufactured wool substitutes are extensively used in the manufacture of woolen and worsted goods. There is no need for the prejudice that is sometimes met regarding these reclaimed materials, for by their use millions of people are warmly and cheaply clothed. If the immense quantity of these materials were wasted, countless persons would be unable to afford proper clothing, as it is difficult to estimate what the price of wool would be; and it is also doubtful if a sufficient quantity could be produced to supply the need. In almost all instances the use of wool substitutes is for the special purpose of producing cloths at a much lower price.

The cloths made from waste products, such as noils, are not much inferior in quality to those produced from the wool from which the noils are obtained; but the great majority of cloths made from other waste products are much inferior. The following are the most important substitutes: noils, shoddy, mungo, extract-wool, and flocks.

Noils are the rejected fibers from the process of combing the different wools and hairs; thus, wool noils are from the sheep, mohair noils from the Angora goat, and alpaca noils from the Peruvian sheep.

[Pg 33] Noils are divided into classes, namely, long-wool noils, short or fine-wool noils, mohair noils, and alpaca noils. They are all obtained in the process of combing, that is, the process which separates the long from the short fibers; the former are known as the "top," and are used in worsted and in the production of mohair and alpaca yarns; while the latter are used to advantage in the production of many different kinds of woolen fabrics. With the exception of length, noils are practically of the same quality as the tops from which they are taken.

Long-wool noils are from the combings of such wools as Leicester and similar wools. These noils, like the wool from which they are obtained, are much coarser in quality and fiber than the short-wool noils. Occasionally, when strength is required in the fabric, these noils are used, and they are also mixed with short-wool noils. Many of the cheviot fabrics are made exclusively of these noils. They are also mixed with shoddy and cotton in the production of dark-colored fabrics, and in medium and low-priced goods requiring a fibrous appearance they are extremely useful.

Short or fine-wool noils are the most valuable, and are obtained from combing Australian and other fine wools. The number and variety of uses to which they are put are innumerable. They are used to advantage in the plain and fancy woolen trade, in the manufacture of shawls and plain woolens of a soft nature, and are also suitable for mixing with cotton in the production of twist threads.

[Pg 34] Mohair and alpaca noils are obtained by the combing of these materials. They are lacking in felting properties, but are lustrous and possess strength, and are most valuable in the manufacture of fabrics where strength and luster are required. These noils are used in the production of yarns for Kidderminster carpets, as yarns for these carpets must possess strength, brightness, and thickness of fiber. They are also used in combination with shoddy and cotton to produce weft or filling yarns for a lower quality of goods.

Shoddy and **Mungo** are in reality wool products, or wool fiber which has previously passed through the processes of manufacture whereby its physical structure has been considerably mutilated. These were first produced about sixty years ago. Shoddy is higher in value than mungo. The value and quality of the waste or rags from which it is made determine the quality or value of the material. Shoddy is derived from waste or rags of pure unmilled woolens, such as flannels, wraps, stockings, and all kinds of soft goods.

Mungo is made from rags of hard or milled character and is much shorter in fiber than shoddy. Its length, varying from one-quarter to three-quarters of an inch, can be regulated by the treatment the rags receive, and by the proper setting of the rollers in the grinding machine. Both shoddy and mungo may be divided into classes. Mungo is divided into two classes, namely, new and old mungo. New mungo is made from rags chiefly composed of tailor's clippings, unused pattern-room clippings, etc. Old mungo is made from cast-off [Pg 35] garments, etc. By a careful selection of the rags previous to grinding, it is possible to make a large number of qualities, and a great variety of colors and shades without dyeing. Owing to their cheapness shoddy and mungo are used in cloths of low and medium qualities. Shoddies are utilized in fabrics of the cheviot class and in the production of backing yarns. Mungoes of the best quality are used in the low fancy tweed trade, in both warp and weft, but chiefly in union and backed fabrics.

Method of Producing Shoddy and Mungo. Before the fibrous mungo is obtained, the rags have to pass through the following necessary preliminary operations:

A. Dusting. This is carried on in a shaking machine, which consists of a cylinder possessing long and strong spikes, which are enclosed, having underneath a grating to allow the dust to pass through. The dust is then driven by a fan into a receptacle provided for that purpose.

B. Sorting. All rags, both old and new, must be sorted, and considerable care must be exercised in this operation, as on this work alone depends the obtaining of different qualities and shades, as well as the securing of the production of a regular and uniform product.

C. Seaming. This is only necessary with the rags procured from garments. It is simply removing the cotton threads from the seams, and any metallic or hard substances from the rags.

D. Oiling. The rags are oiled to soften them and make them more pliable, and thus to facilitate the grinding.

[Pg 36] *E. Grinding.* This is the principal operation, and the rags are made fibrous in this process. The machine by which this is effected is made up of the following parts: feed apron, fluted rollers, swift, and a funnel for conveying the material out of the machine. The principal features of the machine are the swift and its speed. The swift is enclosed in a framework, and is about forty-two inches in diameter and eighteen inches wide, thus possessing a surface area of 2,376 square inches, containing from 12,000 to 14,000 fine strong iron spikes. The speed of the swift may be from 600 to 800 revolutions per minute. The rags are fed by placing them on the traveling feed apron, and are thus conveyed to the fluted rollers. As they emerge from the rollers they are presented to the swift, and by strong iron teeth, moving with exceedingly high surface velocity, they are torn thread from thread and fiber from fiber. The fluted rollers run very slowly, and the rags are held while the swift carries out this operation. By means of the strong current of air created by the high speed of the swift, the mungo is expelled from the machine through the funnel into a specially arranged receptacle. If by any chance the machine should be overcharged, that is, if too many rags are passing through the rollers, the top fluted roller is raised up, and the rags are simply carried, or thrown by the swift, over into a box on the opposite side of the machine without being subjected to the tearing process. The top roller is weighted by levers with weights attached to keep it in position, thus bringing downward pressure to bear upon it, as it is [Pg 37] driven simply by friction. By the adjustment of the feed rollers in relation to the swift, the length of the fiber may be varied to a small degree.

Extract Wool. This is obtained from union cloths, that is, from cloths having a wool weft and warp of cotton, etc., also from cloths having the same material for warp, but possessing a woolen or mungo warp or filling, etc. It is the wool fiber that is required. Therefore the vegetable matter (cotton) must be extracted from it by the process of carbonizing. To effect this, the tissue or rags are steeped in a solution of sulphuric acid and water and then subjected to heat in an enclosed room. The water is evaporated, leaving the acid in a concentrated form, which acts upon the cotton, converting it into powder. The powder readily becomes separated, and thus the cotton is eliminated. The material that is left is well washed to remove all acid, dried, and then passed through a miniature carder, to impart to it the appearance of a woolly and a softer fabric.

Flocks. These are of three kinds, and are waste products of the milling, cropping, and raising operations. The most valuable are those derived from the fulling mill, being clean and of a bright color. They are chiefly used by sail spinners, and in the manufacture of low grade cloths of a cheviot class. White flocks are suitable for blending with wool, and as a rule command a fair price. Raising flocks are those obtained from the dressing or raising gigs, and are applied to purposes similar to those for which fulling flocks are [Pg 38] used. Cutting or cropping flocks are the short fibers which are removed from the cloths in this operation. They are practically of no value to the textile manufacturer, being unfit for yarn production, but are used chiefly by wall-paper manufacturers in producing "flock-papers," which are papers with raised figures resembling cloth, made of poor wool, and attached with a gluey varnish.

CARD ROOM
1. Automatic Feed.
2. Bur Guards.
3. Bur Tray.
4. 1st Top Divider.
5. 2d Top Divider.
6. Workers.
7. Strippers.
8. Doffer Cylinder.
9. Main Cylinders.
10. Main Card Drive on 2d Main Cylinder Shaft.
11. 1st Lickerin.
12. 2d Lickerin.
13. 3d Lickerin.

14. 4th Lickerin.
15. Fancy Hood.

[Pg 39]

CHAPTER IV

WORSTED YARNS

Carding. After the wool is washed it undergoes a number of operations before it is finished into worsted or woolen yarn. [12] The first step in the manufacturing of worsted yarn is to pass the washed wool through a worsted card which consists of a number of cylinders covered with fine wire teeth mounted on a frame. The effect of these cylinders on the wool is to disengage the wool fibers, make them straight, and form a "sliver" or strand. It is now ready for the combing machine.

Combing. The process of combing consists of subjecting the card sliver to the operations of the automatic wool comber, which straightens the fibers and removes all short and tufted pieces of wool. Combing is a guarantee that every fiber of the wool lies perfectly straight, and that all fibers follow one after the other in regular order.

Comb. A comb is a complicated machine. The principal feature is a large metal ring with rows of fine steel pins (pin circles), which is made to revolve horizontally within the machine. By various devices the [Pg 40] wool is fed into the teeth of the ring in the form of tufts. The fibers of the tufts by an intricate process are separated into long and short lengths, and a set of rollers draws each out separately and winds it into a continuous strand called "tops." On leaving the comber, the wool is free from short fibers, specks, and foreign substances, and presents a fine, flowing, and lustrous appearance. The short combed-out wool is called noils, and is used in making carpet yarns, ground up into shoddy stock, or utilized in spinning fancy yarns.

Worsted Tops. American textile manufacturers are finding it advantageous to have their combing done by those who make the work a specialty rather than to do it themselves. In the manufacture of tops all varieties of combing wools are used: Australian, Merino, and Crossbred wools, South American Merino and Crossbred wools, Cape Merino wools, Merino and Crossbred wools grown in the United States, the lustrous wools of pure English blood, Mohair from Asiatic Turkey, and Alpaca from the Andes. Tops are sold to worsted spinneries. [13] Many mills or worsted spinneries send their wools, either sorted or unsorted as they may desire, to a combing mill, where the wool is put into top at a lower price than that at which most spinneries can do their own combing. By means of the naphtha process a larger amount of top from a given amount of wool can be secured than by any other process, and in addition, a top in better condition for drawing and spinning.

[Pg 41]

COMB ROOM

1. Driving pulley on horizontal shaft (2).
3. Boxes containing bevel gears.
4. Pillars.
5. Driving pulley for dabbing motion.
6. Boxes containing dabbing-brush mechanism.
7. Dabbing brushes.

8. Star or stroker wheels.
9. Large circle containing rows of pins.
10. Drawing-off apron and rollers for large circle.
11. Brass boxes or conductors.
12. Guides for comb ball ends.
13. Comb balls (4 ends each).
14. Fluted wooden rollers on which comb balls rest.
15. Comb leg (4 in number).
16. Foundation plate.

[Pg 42] In a strand of combed wool, called top, no single fiber lies across the strand; all lie in the direction of the length. This order is preserved until the fibers have been converted into yarn, which is accomplished by passing through "gill boxes." These gill boxes are machines with bars of iron having upon their surface two rows of minute steel pins, by this means kept perfectly straight. The bars on which they are placed are worked on screws between two sets of rollers. The wool enters between the first set of rollers, and, as it passes through, is caught by one of these gills that is raised up for the purpose, being succeeded by others as the rollers revolve. These gills are moved forward on screws in the direction of the other set of rollers, and the pins in the gills always keep the fiber perfectly straight. The second set of rollers is termed the draught rollers, since by them the wool, after passing through the front rollers, is drawn out and reduced in thickness. This is accomplished because the second rollers revolve at a higher rate of speed than the first rollers, the speed being regulated according to the length of the wool, and the thickness of the yarn to be produced. These gills are used in the production of worsted yarn until the size of the rope of wool has been so reduced and twisted that there is no chance of any fiber getting crossed or out of the order of straightness. A worsted yarn is, consequently, a straight yarn, or a yarn produced from perfectly straight fibers.

[Pg 43]

GILLING

1. Cans containing Comb Ends or Sliver.
2. Balling Head.
3. Stock from Balling Head No. 2.
4. Screws for applying pressure to Back Rollers.
5. Screws for applying pressure to Front Rollers.
6. Faller Screws situated between No. 4 and No. 5.
7. Guard for covering gears which drive Back Rollers.
8. Guard covering gears which drive Balling Head.
9. Balling Head.

[Pg 44] The combing of wool may be dispensed with in some cases, although such a yarn is not in common use. When combing is dispensed with, the gills, in connection with the draught of the rollers, make the fibers straight, and produce a worsted yarn, although such a yarn has a tendency to be uneven and knotty.

Before the wool can be spun it must be made into roving of a suitable thickness. This is done by passing it, after being combed, through a series of operations termed drawing, whose functions are to produce a gradual reduction in thickness at each stage. Although the number of machines varies according to the kind of wool to be treated, still the same principle applies to all.

Spinning. The process of spinning is the last in the formation of yarn or thread, the subsequent operations having for their object the strengthening of the yarn by combining two or more strands and afterward arranging them for weaving or for the purpose for which the yarn is required. It is also the last time that the fibers are mechanically drawn over each other or drafted, and this is invariably done from a single roving. The humidity and temperature of the spinning room must be adjusted to conditions. Each spinner is provided with a wet and dry thermometer so that the best temperature can be ascertained. The most suitable heat and humidity can only be obtained by comparison and observation. A dry and warm atmosphere causes the wool to become charged with electricity and then the fibers repel each other.

[Pg 45]

WORSTED SPINNING. "BRADFORD SYSTEM"
1. Bobbins containing Worsted Yarn.
2. Conical shape caps placed on top of spindles.
3. Tin Wings fastened to Eyeboard.
4. Eyeboard containing pot eyes, through which yarn passes to the bobbin.
5. Scratch fluted front rollers.
6. Leather covered Pressing Rollers for No. 5 Rollers.
7. Smooth metal Pressing Rollers for Back Rollers.
8. Large Front Roller Gear.
9. Pulley for driving Twist gear.
10. Spools of Roving held by a series of pegs.
11. Spindle bands.
12. Sifter plate or rail.

[Pg 46] Worsted yarn is spun by two different methods known respectively as the Bradford or English system and the French system. The difference in these systems of spinning worsteds lies principally in the drawing and spinning processes, a radically different class of machinery being used for each. The combing process is practically the same in both cases, but the wool is combed dry for the French system, and by the English method the stock is thoroughly oiled before being combed. The result of the English method is the production of a smooth level yarn in which the fibers lie nearly parallel to each other. The yarn made according to the French system is somewhat fuzzier and more woolly. On account of the absence of oil, the shrinkage of French spun worsted is considerably less than that made by the Bradford system.

Characteristics of Worsted Yarn. The unique structure of worsted yarn makes it invaluable in the production of textile fabrics in which luster and uniformity of surface are the chief characteristics. The methods by which worsted is formed render it capable of sustaining more tension in proportion to its size than the pure woolen yarn. This feature, combined with its lustrous quality, gives it a pre-eminent position in the manufacture of fine coatings, dress goods, etc. The method of arranging the fibers in the formation of a woolen yarn is such as to produce a strand with a somewhat indefinite and fibrous surface, which destroys to a large degree the clearness of the pattern effect in the woven piece. In the construction of worsted yarn the fibers are arranged in a parallel relationship to each other, resulting in the production of a smooth, hard yarn having a well-defined surface; hence weave-ornamentation of a decided or marked type is possible by its use. There is, in a word, more scope for pattern effects, since the level and regular structure of the yarn imparts a distinction to every part of a woven design. From this peculiarity arises the great variety of effects seen in the worsted dress fabrics, coatings, trouserings, etc., both in colored patterns and in fabrics of one shade throughout.

[Pg 47]

SPOOL ROOM

1. Jack Spooler frame.
2. Drum upon which Jack Spool rests.
3. Jack Spool.
4. Guides for spool ends.
5. Spools containing yarn.
6. Pressers which rest on spools to prevent slack ends.
7. Spool creel.

[Pg 48]

FRENCH SPINNING

1. Balling heads.
2. Bobbins upon which stock is wound.
3. Rub or condenser aprons.
4. Gearing for driving rub motion.
5. Shipper rod and handles.
6. Bobbins held in place in creel by skewers.
7. Weights with system of levers for applying pressure to rollers.

Worsted yarn can be made of pure wool; and as a rule, the wool used in the English system is of fairly good length and uniform staple, for if otherwise it is only with difficulty that the yarn can be spun straight. [Pg 49] Shorter wool can be combed and spun under the French system, and this is the reason why the French system of spinning is being introduced. On the other hand, in the spinning of woolen yarns great length of staple is not essential, for the machinery employed will work the small fibers.

Uses of Worsted Yarn. Worsted yarn may be used in any of the following fabrics:

1. Combed wool yarn for ornamental needlework and knitting, as Berlin, Zephyr, and Saxony wools.

2. Cloth made from combed wool not classified according to material.

a. Fabrics of all wool—serge, bunting, rep, dress goods, with weave effects.

b. Wool and Cotton—union goods, serge linings, lathing.

c. Wool and Silk—rich poplin, pongee, henrietta, bombazine.

d. Alpaca and Mohair—alpaca, mohair dress goods, lusters, braids, laces.

Counts. Yarn is measured by a system of "counts";—the number of yards of yarn to the pound. The counts of worsted yarn are based on the number of hanks in one pound, each hank containing 560 yards. Thus No. 30 worsted yarn consists of 30 hanks of 560 yards each, or 16,800 yards to the pound.

FOOTNOTES:

[12] The distinct difference between worsted and woolen yarns is that worsted yarn is made of fibers that are parallel, while the fibers of woolen yarn run in all directions. The worsted yarn is stronger.

[13] Mills that manufacture worsted yarn.

[Pg 50]

CHAPTER V

WOOLEN YARN

In manufacturing worsted yarn every necessary operation is performed to arrange the wool fibers so that they will lie smoothly and parallel to each other. In the case of woolen yarn every operation is performed so as to have the fibers lie in every direction and to cross and overlap each other.

To produce yarn of the woolen type a set of machinery entirely different from that used in worsted manufacture is necessary. The wool is carded, but no attempt is made to get the fibers parallel. The reduction in thickness of the sliver is not brought about upon the so-called drawing frame, but by a mule frame where the drawing and twisting are done at the same operation. As neither combs nor gills are employed, there is not the same smooth, level yarn, but one which possesses a fringe-like covering or fuzzy appearance that makes the woolen yarn so valuable.

The operation is as follows:

Carding. After washing the material for woolen yarn, it is passed through three carding processes, and from the last of them is taken direct to the spinning frame to be made into yarn. The object of woolen carding is different from carding in any other textile manufacture.

[Pg 51] In most processes of carding the fibers are subjected to a "combing" principle, and the aim is to lay the fibers parallel. Woolen carding aims to open the raw wool fiber, and put it in a perfectly loose condition, without leaning toward any definite arrangement.

The carding machines are called, respectively, first, second, and third breaker. Each machine consists of a complicated series of card-covered cylinders of different sizes, running at different rates of speed — sometimes in the same and sometimes in an opposite direction. These rollers take the wool from one another in regular order until it is finally delivered from the third breaker in a soft, fluffy rope or roll called a sliver. This sliver is wound on a bobbin, and taken from the card to the mule spinning frame.

The sliver on the bobbins from the card is taken to the mule spinning frame where it is passed through rolls, and the sliver attenuated by means of a traveling carriage.

Count. In the case of woolen yarn there are numerous systems for denoting the count, varying with the locality in which it is spun and the character of the product. In the United States there are two systems employed, but the one in most general use is known as "American run counts." This is based on the number of "runs," each containing 1,600 yards to the pound. Thus, a yarn running 8,000 yards to the pound is called a 5 "run" yarn, a yarn with 5,200 yards to the pound is equal to a 3¼ "run."

[Pg 52] In the vicinity of Philadelphia woolen yarn is based on the "cut," each cut consisting of 300 yards, and the count is the number of cuts in a pound. Thus, No. 30 cut yarn consists of 9,000 yards to the pound. No. 15 contains 4,500 yards to the pound.

Woolen yarn is suitable for cloths in which the colorings are blended and the fibers napped, as exemplified in tweed, cheviot, doeskin, broadcloth, beaver, frieze, chinchilla, blanket, and flannel.

[Pg 53]

CHAPTER VI

WEAVING

Preparatory to Weaving. Yarn is wound on bobbins on the ring or mule spinning frame. These bobbins are transferred to a machine called a spooler where the yarn is re-wound on a spool preparatory to making the warp.

A warp is formed by obtaining a definite number of threads (called ends), usually in a precisely designed order of given length, and allowing the ends to wind over a cylinder called a beam. In order to do this it is necessary to have spools placed in a definite position in a frame called a creel.

Before the warp can be placed in the loom so as to weave or interlace it with filling it must be sized. This is necessary for all single twist warp yarns. Its primary object is to increase the strength and smoothness of the thread, thus enabling it to withstand the strain and friction due to the weaving operation. Other objects of sizing are the increase of weight and bulk of the thread and the improvement and feel of the cloth. The warp is usually sized by passing it over a roller and through a bath of a starch mixture. The machine for sizing is called a slasher. The warp is now ready to have the ends drawn in and placed in the loom.

[Pg 54] **Weaving.** Every woven piece of cloth is made up of two distinct systems of threads, known as the warp and filling (weft), which are interlaced with each other to form a fabric. The warp threads run lengthways of the piece of cloth, and the filling runs across from side to side. The manner in which the warp and filling threads interlace with each other is known as the weave. When the word "end" is used in connection with weaving it always signifies the warp thread, while each filling thread is called a pick. The fineness of the cloth is always expressed as so many picks and ends to the inch. The fabrics produced by weaving are named by the manufacturers or merchants who introduce them. Old fabrics are constantly appearing under new names, usually with some slight modification to suit the public taste.

Weaving Processes. In order to understand the different kinds of weaves it is necessary to know, or at least to understand, the process of forming cloth, called weaving. This is done in a machine called a loom. The principal parts of a loom are the frame, the warp-beam, the cloth-roll, the heddles, and their mounting, the reed. The warp-beam is a wooden cylinder back of the loom on which the warp is wound. The threads of the warp extend in parallel order from the warp-beam to the front of the loom, and are attached to the cloth-roll. Each thread or group of threads of the warp passes through an opening (eye) of a heddle. The warp threads are separated by the heddles into two or more groups, each controlled and [Pg 55] automatically drawn up and down by the motion of the heddles. In the case of small patterns the movement of the heddles is controlled by "cams" which move up the heddles by means of a frame called a harness; in larger patterns the heddles are controlled by harness cords attached to a Jacquard machine. Every time the harness (the heddles) moves up or down, an opening (shed) is made between the threads of warp, through which the shuttle is thrown.

A SIMPLE HAND-LOOM
Showing frame, warp beam, cloth-roll, heddles, and reed

The filling thread is wound on a bobbin which is fastened in the shuttle and which permits the yarn to [Pg 56] unwind as it passes to and fro. As fast as each filling thread is interlaced with warp it is pressed close to the previous one by means of a reed which advances toward and recedes from the cloth after each passage of the shuttle. This is done to make the cloth firm. There are various movements on the loom for controlling the tension of the warp, for drawing forward or taking up the cloth as it is produced, and for stopping the loom in the case of breakage of the warp thread or the running out of the filling thread.

Weaving may be performed by hand in hand-looms or by steam-power in power-looms, but the arrangements for both are to a certain extent the same. A great number of different kinds of power-looms are manufactured for producing the various classes of textiles in use at the present time. These looms are distinguished by the name of the material which they are designed to weave, as the ribbon-loom, blanket-loom, burlaps- and sacking-loom, plush-loom, double-cloth loom, rug-loom, fancy cotton-loom, silk-loom, worsted-loom, etc.

Weaving is distinct from knitting, netting, looping, and braiding, which are operations depending on the interlacing of a single thread, or single set of threads, while weaving is done with two distinct and separate sets of threads.

[Pg 57]

MAIN WEAVE ROOM
1. Warp beam containing warp.
2. Lease Rods.
3. Harnesses.
4. Dobby Head motion to lift harnesses.
5. Jacquard Head motion.
6. Cards containing design — working in connection with Jacquard Head motion.
7. Whip Roll.

[Pg 58] **Classes of Weave.** The character of the weave offers the best basis for classification of woven goods. Nearly all the varieties of cloth may be classified from the following weaves:

(1) Plain-weaving,
(2) Twill-weaving,
(3) Satin-weaving,
(4) Figure-weaving,
(5) Double-cloth-weaving,
(6) Pile-weaving,
(7) Gauze-weaving,
(8) Lappet-weaving.

Plain or Homespun Weave. Plain cloth is the simplest cloth that can be woven. In this weave one series of threads (filling) crosses another series (warp) at right angles, passing over one and under one in regular order, thus forming a simple interlacement of the threads. This combination makes a strong and firm cloth, but does not give a close or a heavy fabric, as the threads do not lie as close and compact as they do in other weaves. In plain cloth, if not fulled or shrunk in the finish, the result is a fabric perforated with large or small openings according to the size or twist of yarn used. If heavy or coarse threads are used the perforations will be large; if finer threads, the perforations will be smaller.

This weave is used in the production of sheeting, muslin, lawn, gingham, broadcloth, taffeta, etc. In plain weaving it is possible to produce stripes by the use of bands of colored warp, and checks where both warp and weft are parti-colored. This weave is commonly used when the cloth is intended to be ornamented with printed patterns. In weaving cloth of only one color but one shuttle is used, while for the production of checks, plaids, etc., two or more shuttles are required.

GIRL DRAWING EACH THREAD OF WARP THROUGH THE
REED AND HARNESS READY TO BE PLACED ON LOOM
A, warp beam. B, warp. C, harness frame

Twill Weave. A twill weave has diagonal lines across the cloth. In
this class of weaves the filling yarn or threads pass over 1 and under
2, or over 1 and [Pg 59] under 3, 4, 5, or 6, or over 2 or 3 and under
1, 2, 3, or 4, or over 4 and under 4, 3, 6, etc. Each filling thread does
not pass under and over the same set of warp threads, as this would
not give the desired interlacings. Instead the order of interlacing
moves one thread to the right or left with each filling thread that is
woven. If there are the same number of threads to an inch in warp
and filling, twill lines will form an angle of 45 degrees; if the warps
are closer together than the filling, the angle will be steeper; if the
filling threads are closer together the lines will approach [Pg 60]
more nearly the horizontal. Different effects are obtained in patterns
by variation in the sizes of the yarn and twist, by the use of heavy
threads to form cords, ribs, etc., and by the mixture of vari-colored
materials in the yarn. Often one form of twill-weave is combined
with another to produce a fancy twill-weave. The object of the twill-

62

weaving is to increase the bulk and strength of a fabric, or to ornament it. The disposition of the threads permits the introduction of more material into the cloth, and hence renders it heavier, and of closer construction than in the case of plain-weaving.

Satin Weaves. The object of a satin weave is to distribute the interlacings of the warp and filling in such a manner that no trace of the diagonal (twill) line will be seen on the face of the cloth. In weaving a satin design the filling thread is made to pass under 1 and over 4, 7, 9, 11, or more if a larger plush satin is required. The raising of the warp end to allow the filling to pass under is done in such a way as to prevent twill lines from showing in the cloth. The result is that practically all of the filling is on the face of the cloth. This is called a filling-face satin weave. A warp-face satin weave may be produced by reversing the order; in this case practically all of the warp is brought to the face of the cloth. In this way a worsted warp and a cotton filling might be woven so that practically all of the warp would show on the cloth, and give it the appearance of a worsted fabric. A number of classes of silk goods are made in this way, with a silk filling covering a cotton warp.

[Pg 61] This weave produces an even, close, smooth surface, capable of reflecting the light to the best advantage, and having a lustrous appearance which makes it resemble satin cloth. Satin cloth is made of silk using a satin weave.

Satin weaves are used very largely in producing different styles of cotton and silk fabrics, and are also frequently found in woolen goods. They are more extensively used in the manufacture of damask and table-covers than for any other class of goods. Satin stripes are frequent in madras, shirtings, and fancy dress goods in connection with plain and figured weaves.

Figure Weaving. To produce complicated and irregular patterns in the loom a large number of different openings (sheds) must be made in the warp, and to secure such a large number an attachment is placed on top of the loom called a Jacquard apparatus. The Jacquard is merely an apparatus for automatically selecting warp threads, by which each separate one can be made to move independently of any of the others. It is provided with weighted strings attached to each of the warp threads. The weighted strings are controlled by wire needles which are in turn controlled by perforated cards. Each motion of the loom changes their position and allows some needles to go through the holes in the cards, thus drawing up the warp, while others strike the card and leave the warp down. In this way the perforations of the cards determine the figure of the patterns. The Jacquard is chiefly used to produce patterns of great width in which all or most of the [Pg 62] threads in the pattern move independently. For the weaving of elaborate effects and flowing lines it is practically indispensable. All elaborate designs are classed under the name of Jacquards.

Double Cloth. Double cloth is a descriptive term applied in weaving to fabrics produced by combining two single cloths into one. Each one of these single cloths is constructed with its own systems of warp and filling, the combination being effected in the loom by interlacing some of the warp or filling threads of one cloth into the other cloth at certain intervals, thus fastening them securely together. The reasons for making double-cloths are many. Sometimes it is done to reduce the cost of heavy weight fabrics by using cheaper materials for the cloth forming the back; again it may be to produce double-face fabric; it allows great freedom for the formation of colored patterns which may or may not correspond in pattern on both sides; it is the basis of tubular weaving such as is practised for making pillow cases, pockets, seamless grain bags, etc.; more frequently, the object is to increase the bulk or strength of certain kinds of fabrics, such as heavy overcoatings, cloakings, pile-fabrics, golf-cloth, rich silk, etc.

Pile Weave. A pile weave is a general term under which are classed numerous varieties of cloth woven with a pile surface, as plush, velvet, velveteen, and carpeting of various kinds. Turkish towels are an excellent illustration of pile weaving. A pile surface is a closely set, elastic face covering various kinds of [Pg 63] woolen, silk, and cotton fabrics, and consists of threads standing close together, either in the form of loops or as erect thread-ends sheared off smooth so as to form a uniform and even surface. In the production of a pile fabric a third thread is introduced into the weaving and formed into loops usually by carrying it over the wires laid across the breadth of the cloth. The wires are afterward drawn out, leaving the loops standing; the loops may then be cut so as to form a cut pile, as in velvet and plush, or they may be left in their original form as in Brussels carpet and Turkish towels.

Gauze Weaving. In gauze weaving all the warp threads are not parallel to each other, but are made to intertwist more or less among themselves, thereby favoring the production of light, open fabrics, in which many ornamental lace-like combinations can be obtained. Two sets of warp threads are used, one being the ground warp and the other the "douping," the latter performing the entwining process. Gauze is especially characterized by its openness and yields the lightest and strongest fabric with the least material. When gauze is combined with plain weaving it is styled "leno." Gauze fabrics are designed for women's summer gowns, flounces, window-curtains, etc.

Lappet Weaving. Lappet weaving, really a form of embroidery, is used for producing small designs on cloth by means of needles placed in a sliding-frame, the figures being stitched into the warp. Elaborate figures are beyond the range of lappet weaving, but there [Pg 64] are many small effects that can be economically produced in this manner, such as the detached spots in dotted swiss, and narrow and continuous figures running more or less into stripes. This form of weaving imitates embroidery and is used mainly on plain and gauze fabrics.

[Pg 65]

CHAPTER VII

DYEING AND FINISHING

Dyeing. When a fabric or fiber is impregnated with a uniform color over its whole surface, it is said to be "simply dyed." On the other hand, if distinct patterns or designs in one or more colors have been impressed upon a fabric, it is called printing.

Vegetable dyes were formerly used, but since the coal tar dyes have been discovered the latter are used entirely. Over fourteen thousand colors have been produced from coal tar. Different fibers and fabrics attract dyestuffs with varying degrees of force. Wool and silk attract better than cotton and linen.

Wool Dyeing. The methods of dyeing wool differ considerably from those employed for cotton and other vegetable fibers. They may be divided into three parts: piece dyed, cross dyed, and yarn dyed.

Piece goods are those woven with yarns in their gray or natural state, and then cleansed and dyed in the piece to such colors as are required. They are woven in plain weaves and in a great variety of twills—in fact, in all styles of weaves—and are also made on the Jacquard loom. The principal fabrics in this classification are all wool serges, cheviots, hopsackings, suitings, satines, prunellas, whipcords, melroses, Venetian [Pg 66] broadcloths, zibelines, rain-proof cloths; nun's veiling, canvases, grenadines, albatrosses, crêpes, and French flannels; silk warp Henriettas, voiles, and sublimes. Whenever it is possible, it is better to dye textile fabrics in the form of woven pieces than in the yarn. During the process of weaving it is impossible to avoid getting yarn dirty and somewhat greasy, and the scouring necessary to remove this dirt impairs the color used in weaving. Piece dyeing is the cheapest method of applying color to textiles. The chief fault of piece dyeing is the danger of cloud spots, stains, etc., which do not appear in the other two methods. Then again in the case of thick, closely woven goods the dyestuff does not penetrate into the fabric, and the interior remains nearly white.

The cloth is dyed by means of passing over a roller into a dye vat. Small pieces or "swatches" are taken from the ends of the fabric, and compared with the pattern. For it must be remembered that no two lots of crude dyes are of equal strength, and the wools and cottons of different growths and seasons vary greatly, so that the use of a fixed quantity of dye to a given amount of goods will not always give the exact shade. In comparing a sample with the pattern the two are placed side by side below the eyes (reflected light), and then held up to the light and the eye directed along the surface. A judgment must be formed quickly, as a prolonged gaze fatigues the eye and renders it unable to perceive fine shades of difference.

DYE ROOM
1. Dye tub or vat containing dyestuffs.
2. Rolls or cylinders upon which cloth is wound.
3. Cloth leaving dye tub—being wound upon No. 2 cylinder.

Cross Dyed. Cross-dyed goods may be described [Pg 67] as fabrics woven with black or colored cotton warps and wool or worsted filling and afterwards dyed in the piece. Since cotton has not the same attraction for dyestuffs as wool it is a difficult task to dye a fabric woven with cotton and wool so that both fibers will be identical in depth of color, tone, and brightness. In some cases it is possible to dye a mixed fabric at a single operation, but the usual process is to dye the wool in yarn state and then dye the warp a color as near the color of the wool as possible. In the weaving operation the wool is thrown to the surface. In another method the warp is dyed and woven with a white wool or [Pg 68] worsted filling, and dyed in the piece with a dyestuff that will not affect the cotton. In this way the cotton does not take the wool dye, but retains its original color. This class of work is generally used in mohair, alpaca, and luster fabrics, because the natural brilliancy of the luster wool, alpaca, or mohair in the filling is not impaired as would be the case if the cotton in the goods were subjected to a cotton-dye bath after being woven. The principal cloths in this classification are cotton, warp figured melroses, Florentines, glacés, brilliantines, lusters, alpacas, and mohairs; rainproof cloths, and fancy waistings, and in these may be found the same great variety of weaves and patterns that is to be found in the piece-dyed goods already described.

Yarn Dyed. Yarn-dyed goods differ from those previously described in that they are made of yarns that are dyed before being woven, or yarns spun from dyed wool. Wool may be dyed in the raw state (fleece), slubbing, or yarn. Fleece dyeing is preferable for goods intended to stand friction, and that in spite of wear and tear must preserve their color. It is preferred for dark colored goods where much friction is to be encountered, but is seldom used for light colors, since these would be soiled during subsequent processes of manufacture. In this case every fiber is colored uniformly all over. The yarn from this wool and the cloth woven from it are dyed through and through and do not become grayish or whitish by wear and tear.

Slubbing dyeing is preferred to yarn dyeing, for the dyestuff penetrates the loosely twisted roving, and if [Pg 69] unevenly dyed, the subsequent operations equalize most thoroughly the irregularities in color.

Yarn dyeing is especially applicable to checks, plaids, and suitings, and in their manufacture the drop box loom (a loom with two or more shuttles) is used. Goods manufactured under this classification include cotton warp checks and mixtures; all wool homespuns, mixture coatings and suitings, storm skirtings, rainproof cloths.

These goods are made in a great variety of weaves, the effect in each being secured by the color and the weave.

Piece-dyed fabrics may be distinguished from yarn-dyed fabrics by unraveling threads of each kind. In the case of yarn-dyed fabrics the dyestuff has penetrated through the yarn, while in the case of piece-dyed fabrics the dyestuff has no chance to penetrate as completely as the yarn-dyed fabric.

Textile Printing. Printed fabrics such as print cloths can generally be distinguished by observing the back side of the cloth. If the figure or pattern on the face of the cloth does not penetrate through to the back but only shows the outline, the fabric has been printed. Fabrics are printed by coming into contact with rotating rollers on which the pattern is engraved.

The attraction of cotton for coloring is generally feebler than that of wool or silk. Few of the natural dyestuffs attach themselves permanently without use of a mordant. A mordant is a substance which has an affinity for, or which can penetrate, the fiber to be colored, and which possesses the power of combining [Pg 70] with the dyestuff and thus forming an insoluble compound upon the fiber. Cotton is dyed in an unspun state, and also as yarn or spun thread, either in the hank or skein. Silk is dyed in unspun skeins, although to a considerable extent it is also dyed in the piece.

Styles. Since styles and designs are constantly changing it is necessary for the mills to meet this demand by producing new styles. Some of the patterns which are at this time considered to be in the best style would have appeared much out of date two or three years ago, while perhaps a few years hence, the patterns which are now almost obsolete will, with some changes, become the most popular sellers of the season. As the mill officials or designers are not out among the trade, they are not in a position to judge what lines or patterns would most likely appeal to the market. This information is obtained by the "styler" of the selling house. The styler receives all the latest foreign samples and fashion papers from abroad, and often goes or sends his representative to Europe to ascertain what goods, designs, and colors are taking well over there. The selling agent or styler then supplies the designing department of the mill with all the samples, information, and suggestions necessary in getting out the samples.

Construction of Cloth. In reproducing a sample of cloth in the mill it is necessary that the construction of the cloth be first known, that is, there must be ascertained the width, warp ends, and picks per inch, the number or size of the yarn used for the warp, the number that is used for the filling, and the number of [Pg 71] ounces per yard or yards per pound. Then the interlacings of the threads in the sample must be picked out in order to get the design or weave on the design paper, from which the data are obtained for regulating the movement of the harness or heddles. Design paper is paper ruled by lines into a number of squares. An imitation of the cloth can be produced on this paper by showing the interlacings of the warp and filling. This is done by filling in certain squares with paint, or pencil marks, while the others are left empty. In practical work it is the general custom to make a cross with a pencil to indicate the squares that are to be filled in, thus showing that the warp thread is over the filling thread at this point. When a square is left blank it shows that the warp thread is under the filling at this point. When a warp thread is up on a certain pick, the harness which controls this thread must be raised on this pick.

Finishing. The fabric as it comes from the loom is in an imperfect condition for use. When worsted fabrics leave the loom they require but few and simple finishing operations, and in this respect differ much from woolen cloths, which require elaborate finishing operations. The finishing processes of woolen and worsted cloths are similar. The following description of processes and machines gives a clear idea of the necessary finishing processes for a standard woolen or worsted cloth; for particular styles of finish the processes must be varied in accordance with the particular requirements of the style of fabric in hand.

[Pg 72] **Perching.** The fabric as it comes from the loom receives a perching and measuring inspection at the weave room before leaving for the finishing room. This examination is to detect quickly such imperfections as require prompt attention at the loom.

Burling. Every knot that has been tied in the threads during winding, dressing, beaming, and weaving, must be looked for and felt for during burling, carefully drawn to the surface of the cloth, and then clipped off with the scissors, leaving the ends long enough so that no space without a thread will occur. Threads which are found loose on the face or back of cloth, caused by the weaver having tied in a broken end, should be cut off and not pulled off. All places where threads are not woven in are marked so that the sewing-in girl (mender) can adjust such places. The cloth is subjected to perching again. It is examined for imperfections, and when these are found, they are marked with chalk to call the attention of the menders to such places.

Mending. The object of darning or mending is to make all repairs in the structure of the cloth before the process of fulling. The mender must have a good eye for colors necessary to produce various effects and for the interlacing of the threads. More exact work is required for threadbare fabrics that require little if any finishing afterward, than in dealing with a face finish fabric, where the nap is to be raised and will cover many imperfections so that they will never be noticed in the finished cloth.

[Pg 73] **Fulling.** The object of fulling is to render woolen and worsted goods stronger and firmer in body. Fulling is similar to felting, the principal object of each being to condense the fibers, thereby increasing the firmness. Certain varieties of woolens are fulled nearly one-half their original width and length. The process of fulling includes three steps: cleansing, scouring, and condensing the fibers of the cloth. The object of scouring is to get rid of oil used preparatory to spinning, and to remove from the cloth stains and the sizing used in dressing the warp. The cloth is first saturated with hot water and soap, and is then scoured and rubbed between the slow-revolving rollers of the machine from two to eighteen hours, according to the character of goods and the amount of shrinkage desired. The more prolonged the operation, the more the material shrinks. When sufficiently fulled, the length of cloth is scoured to free it from soap. This is done with water, warmed at first, but gradually cooled, until at the end the cloth is worked in cold water. Next the cloth is stretched uniformly in all directions, so that it may dry evenly without wrinkles or curls. Sometimes the cloth is placed in a hot-air chamber to hasten the drying. The fulling or shrinking is effected by the application of moisture, heat, and pressure. Every one is familiar with the fact that woolen blankets, flannels, and hosiery tend to contract with frequent washings, gaining in thickness and solidity what they lose in elasticity. Such shrinkage is greatly hastened when they are rubbed vigorously in hot water and then allowed to cool suddenly. [Pg 74] This change is due to the physical properties of the wool fiber.

Such goods as beavers, kerseys, meltons, and fancy cassimeres are seldom fulled more than one-sixth of their woven width, while worsted goods are shrunk but a small fraction of their woven width. The amount of fulling received is the distinguishing feature of many varieties of cloth. In the treatment of broadcloth, doeskin, and all nap finished woolens, the fulling is carried to a point where the fibers become densely matted, obliterating all traces of the weave and giving the cloth the appearance of felt.

Crabbing. After the cloth has been dried in the hydro extractor, where it throws off superfluous moisture, it must be stretched full width for the future finishing processes, and "set" at this width.

Crabbing consists of two operations, first the loosening process, then the setting process. Goods are run on a cylinder, then passed over several rolls, and are kept tight so as to avoid wrinkles. The cylinders are immersed in hot water and the goods are allowed to rotate in this water for about twenty minutes, after which they are taken out for one or two hours. They are then returned to the machine for about twenty-five minutes and are subjected to boiling and also to additional pressure. The boiling water sets the fabric and the additional pressure gives the desired finish.

Tentering. The object of tentering [14] is to straighten and level the fabric. After the cloth leaves the [Pg 75] tentering machine it has lost its natural moisture, and is not at all fitted, as far as fiber condition is concerned, for the napping. To bring it into a fit state for this operation it is passed through a trough containing a brush which gives it the desired moisture. It is then ready for napping.

Napping. Most cloths at this stage of finishing are more or less unsightly on account of long and irregular fibers on the surface. A nap may be raised on the surface of a fabric for various reasons: in order to render the material warmer, softer, or more pleasant to the touch, as in the case of blankets and flannels intended to be worn next to the body; or for the purposes of increasing the durability of the fabric, as in the case of melton, kersey, broadcloth, and similar goods; or a nap may be raised with a view to removing all the fiber from the underlying structure in order to leave the pattern of the cloth well defined and free from hairiness. The covering of nap over the surface of the fabric tends to conceal many defects caused by imperfect yarns and faulty weaving. Coarse, inferior yarns at best produce an unsightly fabric, but when the cloth constructed of such threads is finished with a fine, delicate nap the surface takes on a softer and richer appearance. Not only are the defects in the structure concealed, but the material is rendered more sightly and desirable and appears to be more expensive than it really is.

The operation of napping is performed by passing the cloth in a tightly stretched condition over a revolving [Pg 76] cylinder covered with teasels or steel hooks. These thousands of little hooks scratch the entire surface of the cloth, opening up the short fibers and covering the whole with a nap. Since the fibers are of different lengths it is necessary to brush the fabric vigorously and then pass it through the shearing machine in order to make an even and uniform length. The shearing machine acts on the principle of the lawn mower and either cuts the nap completely or leaves a pile surface. The cloth is cleaned by passing through a brushing machine.

Pressing Machine. The fabric now requires consolidating and lustering, or "smarting up" in appearance—practically pressing—before it is forwarded to the warehouse. This is done by passing the cloth over a pressing roll heated to a high temperature. Having obtained a satisfactory luster, it is necessary to fix this by winding the cloth on rollers and allowing dry steam to pass through the piece. This fixes a permanent luster and finish on the piece and sets it so as to prevent shrinkage. The cloth is now packed and sent to the jobbers or tailors to be cut up into suits.

Theories of Coloring in Textile Design. The three primary elements of textile design are weave, combination of form, and blend of colors. They enter either separately or in connection with each other into every species of loom effect. Weave relates specifically to the build or structure of the cloth and is an indispensable factor in any type of cloth. Schemes of weaves will produce in one operation an even and firm cloth, decorated with a type of pattern that usually consists [Pg 77] of minute parts but which is pronounced and decided in combination. Combination of forms is a surface decoration obtained by uniting straight and curved lines. Color brightens and improves the qualities of the design. In fact, the discarding of color shades would diminish the elegance of the design and impoverish its appearance and would practically destroy the woolen industry. Whether the pattern be stripe, check, figure, or intermingled effect, it obtains its outline and detail from methods of coloring adopted. In worsted there is a larger diversity of weave design than in woolen; but still colors are very extensively employed to develop effects due to weave and form, and also to impart a cheerful and lustrous appearance to cloth.

Patterns in dress fabrics, shirtings, and other articles made entirely of cotton are frequently mere combinations of fancy shades, while fabrics composed of silk and jute materials, including silk ties, handkerchiefs, etc.—in fact the cloths in which fancy shades are used—show that coloring and its combinations in all woven product embellished with design, are elements which give tone and character to the styles. Though the cloth may be soft to the touch, substantially made, of uniform structure, and skilfully finished, yet a lack of brightness and elegance in coloring so powerfully detracts from the appearance of the pattern that these qualities alone are not sufficient.

On subjecting cotton, silk, wool, and worsted goods to inspection, color is found to have a different tone or cast in each fabric. Fancy colors in cotton, while [Pg 78] decidedly firm and clear in effect, are non-lustrous, raw, and dull in toning. Silk colorings, on the contrary, possess both compactness and brilliancy; woolen colorings have a unique depth and saturation of hue characteristic of the material employed in the manufacture of woolen goods; while worsted colorings are bright, definite, and smart in appearance.

These differences are due to the physical properties of the several fibers. Thus a filament of silk is transparent and shines like smooth glass when light falls upon it; that of wool is solid and opaque in the center, but its exterior consists of a multitude of semi-transparent scales which, when of large dimensions and uniformly arranged—

as in the best qualities of wool—reflect light with a small amount of dispersion and impart to the woven material a lustrous aspect. Cotton has no such partially transparent sheath. What light is reflected is so broken up that the color is poor. Compare three plain woven crimson textures made of silk, wool, and cotton respectively. The first literally shines; luster, brilliance, and richness are the elements of its coloring. Though bright, it lacks that fulness and depth of color which belongs to the wool product, whose millions of filaments, closely compounded, all tinted alike, possess a peculiar bloom and weight of color not to be found either in the silk or cotton article. Lastly, take the crimson calico. How deficient in warmth and richness it seems to be, after examining the woolen and silk texture! It is dull and has a raw and deficient character.

[Pg 79] The various methods of employing fancy shades in patterns obtained in the loom may be briefly summarized:

I. In mixture cloths, for suitings, coatings, etc.

a. By combining or blending various colors of materials.

b. By combining several classes of twist threads.

II. In plain, twilled, mat, and fancy weave designs for trouserings, coatings, suitings, jackets, dresses, costumes, flannels, shirtings, etc.

a. By introducing colors into the warp, forming stripes.

b. By introducing colors into the filling, producing spotted patterns.

c. By introducing colors into both warp and filling, giving checks, broken styles, etc.

III. In figured designs for dresses, vestings, etc.

a. By using one or several series of extra warp yarn.

b. By using one or several series of extra filling.

Dress goods fall naturally into two distinct classes when regarded from the standpoint of fashion—staples and fancies. Staples are those fabrics which are made of the same construction year in and year out. They vary only in coloring to meet the changes of fashion.

The Staples are:

Brilliantines,
Sicilians,
Mohairs,
Imperial Serges,
Storm Serge,
Cheviots,
Panamas,
Batistes,
Taffetas,
Voile,
Nun's Veiling,
Cashmere,
Shepherd Checks.

[Pg 80] The Fancies are:

Produced through
Variation of weave,
Variation of color,
Variation of color and weave:
Brocades,
Cuspettes,
Meliores,
Hopsacking, etc.
Coloring includes:
Stripes,
Checks,
Plaids,
Malenges,
Mixtures.

Prior to the factory era our fathers and mothers made homespun clothes and wore them till they had passed their period of usefulness. The average consumption of wool at that time averaged not more than three pounds per capita. As wealth increased the home loom and spinning-wheel were slowly supplanted by the mill and factory. The different textile manufacturers at length found that competition was so keen that it was necessary to adulterate, particularly any fabric that was popular. The classes of goods that are most adulterated are the expensive fabrics, those of wool and silk. There are such changes of fashion in dress at the present day that garments composed of materials formerly considered good enough are often thrown aside as old-fashioned when only half worn. Manufacturers cater to the whims and fancies of people and import to this country foreign styles. The rapidly changing styles cause people to throw upon the market a great amount of cast-off clothing only partially worn.

[Pg 81]

SALES DEPARTMENT OF THE SELLING AGENT OF A LARGE MILL

[Pg 82] The result is that there is not wool enough to provide the public with clothing made of new wool. The requirement per capita has risen to six pounds. The immense amount of fiber in cast-off clothing does not find its way into the paper mills, but rather into the shoddy mill, where it is remanufactured into cloth again, or where part of the fiber is mixed with good wool to make "pure wool" cloth. In other words, the rapidly changing styles of to-day and the limited supply of wool are responsible for the wholesale adulteration which is being practised in modern cloth manufacture. This adulteration furthermore is becoming more and more difficult to detect by reason of the rapid improvements made in the finishing processes of cloth manufacture. Hence the necessity for people to know how and why adulteration occurs, how it affects prices, and what are the means of detecting it. Shoddy is considered a legitimate adulteration in woolen and worsted goods. The following adulterations are not legitimate unless sold as such:

1. Cotton combed with wool.

2. Thin cotton threads twisted in with worsted during the process of drawing.

3. Cotton threads of the same color as the wool or worsted used as filling or warp.

4. Cotton veneered with wool.

5. Cotton threads of the same color as wool used in weaving.

FOOTNOTE:

[14] Tentering is carried on in the English mills.

[Pg 83]

CHAPTER VIII

WOOLEN AND WORSTED FABRICS [15]

Albatross. A dress fabric of worsted warp and worsted filling; of open texture and fancy weaves.

Alpaca. A thin fabric of close texture made from the fibers of an animal of the llama species; mixed with silk or with cotton. It is usually woven with cotton warp and mohair filling. Imitations of all cotton are manufactured and sold under this name.

Corded Alpaca. Corded weave, lengthwise of the piece, cotton warp alpaca filling; one of the first products of the American loom.

Angora. The fiber of this goat is commercially known as mohair. The skins are largely used in the making of children's muffs, for the scalps of dolls, and for trimming coats and capes. Carriage robes also claim a good share of the skins; the hair, being nearly one foot in length, makes them beautiful and serviceable. The fiber enters largely into that class of goods known as Astrakhan, Crepons, Plushes, Brilliantines, Zibelines, fine Cashmeres, and many other fabrics usually sold as all wool or worsted, according to the [Pg 84] mode of preparing the stock before spinning into yarn. It is found in the finest of silk and worsted fabrics for ladies' wear, also in linings, mittens, and fine cloaking and overcoating. It is noted especially for its water repelling qualities, its beauty, and high luster; and not so much for its warmth-retaining properties, for which wool stands unequalled.

Astrakhan. A fabric manufactured from Astrakhan fiber; of a curly, wavy surface applied to a curly faced cloth resembling Astrakhan fleece.

Bandanna. From the Indian *bandanna*, to bind or tie. In dyeing, the cloth is tied in knots when dipped, and thus has a clouded effect.

Beaver. A heavy cloth manufactured of fine wool, with a finish on the surface to resemble the fur of the animal by that name.

Fur Beaver. Similar in many respects to Beaver, but having on its surface a long, dense nap, in imitation of the fur of the Beaver. Used for overcoats, cloaks, and capes.

Bedford Cord. A fine woolen fabric, with fine recesses running with the piece, and extensively used for ladies' dress goods. An all wool cloth of close texture for gentlemen's clothing. The recesses may also be made with fine cotton yarn hidden in the wool filling.

Beige. Cloth of undyed or natural wool. The name is the French word for "natural."

Bindings. A species of narrow fabric of silk, worsted or cotton, for binding the edges of garments, the bottom of dress skirts, etc.

[Pg 85] **Bombazine.** A twilled fabric of which the warp is silk and the filling is worsted.

Bottany. A term applied to worsted yarns made from bottany wool. It is considered the finest of all worsted yarns, and is used for fine fabrics of close texture.

Boucle. Curled hair or wool woven in any cloth in such a way as to show the curl makes boucle. The word is French for curl.

Broadcloth. Broadcloth is a soft, closely woven material with a satin finish. The best qualities are called satin broadcloth.

The origin of broadcloth dates back to early times, the first historical mention of it being made in 1641. In America, among the first products manufactured by the colonial woolen mills were black and colored broadcloths, and these (with satinets) formed the distinctive character of American woolen fabrics at that time. They were honestly made of pure, fine-fibered Saxony wool, and sold as high as $6.50 per yard.

The warp and filling are made of carded wool so that the web (cloth) will shrink or full evenly. The stock is generally dyed in the raw state when used for men's wear. When taken from the loom it does not have the smooth, lustrous appearance which is its distinctive feature. It is rough and dull colored, with the threads showing plainly. To improve its appearance it is first subjected to the action of the fulling mill, with the result that the fibers of the warp and weft become entangled to such an extent that the cloth never unravels. Then the cloth is slightly napped and sheared down close, [Pg 86] in order to produce a smooth, even surface. Next it is successively wetted, steamed, calendered, and hot pressed for the purpose of bringing out the luster. It is commonly twill woven, but is sometimes plain, finished with a slightly napped and lustrous face. It must have a bright, beaver finish, and be close and felty in the weave.

The broadcloth used for women's clothing is of a lighter weight and is generally piece dyed. It is used for ladies' suits, coats, and gentlemen's evening dress suits, frock coats, and tuxedos. It is expensive; prices range from $1.75 to $3.50 per yard in ladies' broadcloth, and higher for men. The price depends on the quality of wool used, and uniformity of the nap and perfection of the finish.

Bunting. A plain even thread weave of mohair, wool, or worsted, used mostly for making flags. The name is from German, *bunt*, meaning variegated or gay colored.

Caniche. A name given to curled wool fabric showing the effect of the coat of the caniche, a French dog.

Cashmere. A cloth made from the hair of the Cashmere goat. The face of the fabric is twilled, the twills being uneven and irregular because of the unevenness of the yarn. Cashmere yarn was first hand spun. The goats are grown for their wool in the vale of Cashmere in the Himalaya Mountains.

All Wool Cashmere. As no material by this name exists there can be no definition. When the term is used in defining a fabric, it is a delusion and a snare.

[Pg 87] **Cashmere Double.** A cloth having Cashmere twill on one side or face and poplin cord on the reverse.

Cassimere. The name is a variation of Cashmere. Cassimere, when properly made, is of Cashmere wool. Usually a twill weave.

Castor. Same as beaver, of a light weight.

Challis. (Also spelled *challie*.) A name given to a superior dress fabric of silk and wool first manufactured at Norwich, England, in 1832. In texture the original material was soft, thin, fine, and finished without gloss. When first introduced it ranked among the best and most elegant silk and wool textures manufactured. It was composed of fine materials, and instead of giving it a glossy surface, such as is usually produced from silk and fine wool, the object was to make it without luster. The name is now applied to an extremely light weight summer dress fabric, composed of either cotton or wool, or a mixture of these fabrics. In structure it is both plain woven and figured, the ornamental patterns being produced either in the loom or yarn, dyed or printed. It is not sized. All wool challis does not differ essentially from the old-fashioned muslin delaine. Most challis patterns are copied from the French silks, and this accounts in part for their tasteful designs and artistic effects. French challis is a material similar to the above, though usually characterized by a more glossy finish.

Cheviot. A descriptive term of somewhat loose application, being used indiscriminately of late years to denote almost any sort of stout woolen cloth finished [Pg 88] with a rough and shaggy surface. Originally the fabric known as cheviot was woven in England, from the strong, coarse wool of the Cheviot sheep, whence the name.

It is at present a worsted or woolen fabric made of cheviot or "pulled wool," slightly felted, with a short even nap on the surface and a supple feel. Worsted cheviots, in plain colorings or of fancy effects, are manufactured from combed yarn. Woolen cheviots are made from carded yarn. The greater portion of this class of goods in carded yarns contains little or no new wool in its make-up. Shoddy, mungo, and a liberal mixture of cotton to hold it together, blended in the many colorings, help to cover the deception. Prices range from 50 cents to $3.00. The material is plain or twill woven, and has many of the qualities of serge.

The distinguishing feature of cheviot, whatever the grade of cloth, is the finish, of which there are two kinds. One is known as the "rough" finish, and the other as the "close" finish. Real cheviot is a rough-finished fabric, composed of a strong, coarse wool and fulled to a considerable degree. The process of finishing cheviot is simple, and practically the same methods are followed for both the "rough" and the "close" styles. On leaving the loom the cloth is first washed in soap and water to remove any dirt or other foreign matter it may contain. It is then fulled, which consists in shrinking the cloth both in length and breadth, thus rendering the texture heavier and denser. Next it is "gigged" or napped. This is accomplished by passing the face of the matted cloth against a cylinder covered with sharp [Pg 89] pointed teasels which draw out the fibers from the yarn. This operation is continued until a nap more or less dense is raised over the entire surface.

From the gig the cloth is taken to the shearing machine, the revolving blades of which cut the long, irregular nap down to a uniform level. Sometimes the style of finish called for is that approaching a threadbare cassimere, and in this case great care is necessary to prevent the blades from cutting the yarn. In the rough finish the nap, although sparingly raised, is comparatively long. Having been napped and sheared, the cloth is pressed and carefully examined for defects, then brushed, pressed, and highly steamed. When measured, rolled, and steamed, it is ready for market, and is used mostly for ladies' and gentlemen's suitings. The pattern and design are light stripes and checks of small dimensions. Cheviot is a name given to many materials used for suiting.

Chinchilla. Heavy coating with rough wavy face. The name is Spanish for a fur-bearing animal of the mink species.

Chudah. Applied to billiard cloth; relates to color. Chudah is the Hindoo name of a bright green cloth.

Corduroy. Heavy corded cotton material used for servants' livery. The name is from the French *Corde du Roi*—king's cords.

Côte Cheval. In France corded cloth for riding costumes, such as Bedford cord, is called côte cheval, the application being through *cheval*, horse; *côte*, ribbed or lined.

[Pg 90] **Coupure.** Coupure is French for cut through. Coupure or cut cashmere is a cashmere weave showing lines cut through the twills lengthwise of the piece.

Covert. Heavy twilled cloth in natural undyed shades, used in England for men's overcoats worn while riding to covert in fox hunting.

Delaine. From the French "of wool"; applies to the most primitive weave of plain wool yarn. Thirty years ago delaine was the staple dress goods stock. It was made in solid colors.

Diagonal Cheviot. Same as cheviot, only in the weaving the pattern is marked by zigzag lines or stripes.

Doeskin. Of the broadcloth range, made with shiny napped face, soft finish, as the pelt of a doe.

Drap d'Été. A heavy cashmere or double warp merino, with the back teasled or scratched, used mostly for clergymen's clothing and in lighter weights for women's dresses. The name is French for "cloth of summer."

Empress Cloth. Similar to poplin; made of hard twisted worsted filling and cotton warp. Was made a success in the early seventies of the last century by the Empress Eugenie of France. Empress cloth was a staple in all well-regulated dress goods lines.

Épingline. A fine corded fabric of wool or silk, showing the cords woven close together and appearing as if lined with a pin point. This application is from *épingle*, French for pin.

[Pg 91] **Etamine.** French name for bolting or sifting cloth, made of silk for sifting flour; applied to mesh or net weaves in America.

Felt. Fabric made by rolling or pressing a pulpy mass or mixture of wool into a flat mat. The name is from the process. To felt is to mix and press into shape.

Flannel. Wales appears to have been the original home of flannel, and history informs us that this was the only textile produced in that country for hundreds of years. It is constructed either of cotton or wool, or of an intermixture of these fibers, and is a coarse-threaded, loosely woven, light-weight fabric, more or less spongy and elastic, with an unfinished, lusterless surface. Generally speaking all grades of plain colored flannel are piece dyed, the soft open texture of the goods permitting the fibers to absorb the dye as readily in the web as in the yarn. Flannels are subjected to several finishing operations, such as fulling, teaseling, pressing, and stretching. Flannels do not require a great deal of fulling. All that is necessary is enough to give a degree of stability and body to the goods.

Dress Flannel. All wool fabric used chiefly for women's winter dresses; also called flannel suiting. It has a diversity of qualities, colors, and styles of finish. It is commonly put up in double fold, width from twenty-six to fifty inches.

French Flannel. A fine, soft twill, woven variety dyed in solid shades, and also printed with patterns after the manner of calico; used for morning gowns, dressing sacques, waists, etc.

[Pg 92] **Shaker Flannel.** A variety of white flannel finished with considerable nap, composed of cotton warp and woolen weft.

Indigo Blue. A superior all wool grade used in the manufacture of men's suits and particularly for the uniform of members of the G. A. R.

Mackinaw. The name applied to an extra heavy blanket-like material used in cold climates by miners and lumbermen for shirts and underwear.

Navy Twilled Flannel. A heavy all wool variety commonly dyed indigo blue, commonly used in the manufacture of overshirts for out-door laborers, firemen, sailors, and miners.

Silk Warp Flannel. A high grade, pure variety of flannel woven with a silk warp and a fine woolen weft. It is a very soft, light-weight, loosely woven flannel and runs only in narrow widths, twenty-seven inches. If the finishing process is carried beyond full-ing the texture is rendered hard and firm, the cloth thus losing its softness and elasticity. In the teaseling process it is necessary for the nap to be raised only slightly, and this is commonly done in the direction of the grain or twist of the warp. The perfection of a flannel finish lies not in the smooth appearance of the cloth, but in its full, rich softness. Sometimes the nap is sheared, but more often it is pressed down flat upon the face of the cloth. After a thorough drying, and careful examination for defects, the goods are rolled on boards, and are ready for market. It is used for infants' wear and shawls, for undergarments, bed coverings, and also [Pg 93] to some extent for outer garments in weights and styles adapted for that purpose.

Baby Flannel. A very light-weight variety woven of fine, soft wool, smooth finish, bleached pure white.

Florentine. A heavy twilled mohair fabric for men's wear which is sold largely to Italy and Spain. The name is from Florence, Italy.

Foule. A twilled, unsheared cloth; that is, the face appears to be unsinged, and shows the woolly roughness in a slight degree. The cloth when woven in the gray is fulled or shrunken in width by soaking in soapsuds and passing it while wet through holes of different sizes in a steel plate. The name is from *fouler*, French, to full or shrink.

Frieze. Frieze is a coarse, heavy cloth with a curly surface and made at first of lamb's wool. It is now made from coarse grades of wool. It is thick and heavily napped, and is used in the manufacture of warm outer garments, particularly for men's wear. It was named after the people of Friesland in Holland in the 13th century, and is famous to-day as an Irish fabric. Irish frieze has extraordinary durability, and the fibers are the longest and strongest made. The weave is plain, small twill, or herring bone. When not of a solid color it is usually a mixture, the colors being mixed in the raw state. The wool is dyed in the raw state in mass, then doubled after spinning.

Gloria. Plain weave of silk and wool, and silk and cotton; first made for umbrella covering. Name means bright.

[Pg 94] **Granada.** Popular weave of mohair, made in coating weight for Spanish trade. Granada is a city in Spain.

Grenadine. Originally a plain, openwork, net-like fabric of silk, mohair, cotton, or wool. We have grenadines in Jacquards and in set patterns. The name is an adaptation of Granada.

Henrietta Cloth. A twilled cashmere of light weight and high finish, originally made with silk warp and wool filling in Yorkshire, England. The name was given in honor of Henrietta Maria of England, Queen of Charles I. The silk warp, hand-woven fabric was first produced about the year 1660.

Homespun. A rough, loosely woven material made from coarse yarn. It is soft but rather clumsy. A general term used to designate cloth spun or wrought at home. The homespun of the present day is a woolen fabric in imitation of those fabrics made by hand before the introduction of textile machinery. It is made of a coarse, rough, and uneven thread; usually of plain weave and no felting. It was woven by the early settlers of the Eastern and Southern States. It is now used as woolen suiting for men's wear and in various kinds of coarse, spongy, shaggy cloth for women's gowns.

Hop Sacking. A coarse bagging made commonly of a combination of hemp and jute, used for holding hops during transportation. The name hop sacking is also applied to a variety of woolen dress goods made from different classes of yarn. It is made of carded woolen fabric of the plainest kind. The cloth is [Pg 95] characterized by an open weave, and a square check-like mesh, the structure being designed to imitate that of the coarse jute bagging. It has very little finish, is usually dyed in solid colors, and is used for women's and children's dresses.

Jeans. Cotton or woolen coarse twilled fabric. In cotton used for linings, in wool for men's cheap clothing. The name is from a Genoese coin, relating to the price of the cloth; so much for one jean.

Kersey. A very heavy, felted, satin finish woolen cloth made with the cotton weave or cross twill for face, and cotton weave or four harness satin for back. It was originally made with fine Merino lamb's wool for face, and somewhat coarser grade for back. The cheaper grades are manufactured from a fine-fibered wool and shoddy, with low grades of shoddy and mungo for back. It is named from an English town, Kersey, where from the eleventh to the fifteenth century a large woolen trade was carried on. The Kersey of early history was a coarse cloth, known under different names, and before knitting was used for stockings. In the construction of Kersey the cloth is woven a few inches wider in the loom (and correspondingly longer) than it is to appear in the finished state. This is done in order that the meshes may be closed up in the fulling mill to insure a covering of threads. Previous to fulling, however, the face of the cloth is gigged to produce a good covering for the threads by forming a light nap, which is fitted in. In the fulling operation, which comes next, the cloth is shrunk to its proper width and density, usually to a degree rendering [Pg 96] it difficult to see the individual warp and filling threads, so closely are they matted together. Fulling is followed by gigging, and in this process a nap more or less heavy is raised on the face of the goods by means of teasels. The cloth is run through the gig several times and then sheared in order to render the fibers forming the nap short, even, and of uniform length. Great care is exercised in the shearing, as the nap must be cropped quite close and yet not expose the threads or cut the face. The next operation is scouring or steaming, in which live steam is forced through every part of the goods for the purpose of developing the natural luster of the wool. In case the goods are to be piece dyed, the dyeing follows scouring. After steaming, the cloth is thoroughly matted and gigged again, care being taken to avoid stirring up the ground nap. It is then dried and the nap brisk-ly brushed in a steam brusher and laid evenly in one direction. Again the cloth is slightly steamed and primed, face up. The result of this treatment is the production of a texture firm, yet pliable, with a highly lustrous face and one not liable to wear rough or thread-bare. Kersey is used for overcoats.

Kerseymere. Light weight twilled worsted; same derivative of name as Kersey.

Linsey Woolsey. Coarse cloth of linen and wool used as skirtings by the British peasantry. The name is from the components of the cloth.

Melrose. Double twilled silk and wool fabric; named for Melrose, a town on the Tweed, in Scotland.

Melton. A thick, heavy woolen fabric with short [Pg 97] nap, feel-ing somewhat rough. Meltons are made firm in the loom. The weaves for single cloth meltons are usually plain, and three or four harness twill. For double cloths the plain weave is used, or a weave with a plain face and a one-third weave on the back. All trace of the weave is destroyed in the finishing. The colors usually black or dark blue.

Meltonette. A cloth of the same general appearance as melton, of light weight, for women's wear.

Merino. A fabric woven of the wool of the Merino sheep, twilled on both sides, the twill being uneven. Merino resembles cashmere.

Mohair Brilliantine. A dress fabric resembling alpaca, of superior quality, and sometimes finished on both sides. The name is from the Arabic *mukayyan*, cloth of goat's hair. It is made from the long, silky hair of the Angora goat of Asia Minor, a species which is being introduced into the United States. The fabric has a hard, wiry feel, and if made from the pure material has a high luster. It has cotton warp and luster worsted filling. The weave is plain ground, or with a small Jacquard figure, and when a very lustrous fabric is wanted, the warp yarn is of finer counts than the filling yarn. The warp and filling yarns are dyed previous to weaving. They may be of the same color or different colors. The contrast of colors in connection with the weave gives the fabric a pretty effect. Fabrics made with dyed yarns are usually given a dry finish, that is, simply run through the press and cylinder heated, after which they are rolled and then packed. Those made [Pg 98] with undyed filling are first scoured, then dyed, after which they are run through a rotary press with fifty or sixty pounds of steam heat. Mohair brilliantine is used for dress goods.

Montagnac is heavy overcoating. The French *montagne*, for mountain, is the origin of the name, being for mountain wear.

Orleans. Cloth of cotton warp and bright wool fulling, made in Orleans, France. Many of the so-called alpacas and mohairs of to-day are Orleans. These fabrics are mostly cross-dyed, that is, fabrics with warp and filling of different shades. After weaving they are cross-dyed or redyed to give solid colors and glacé effects.

Panama Cloth is a plain weave worsted fabric of no uniform construction or finish. Fabrics sold under this name vary considerably. They are of solid colors, usually piece dyed, and are used for suitings.

Prunella. From the French *prunelle*, which means plum, a stout worsted material named from its color, which is a purplish shade similar to that of a ripe plum. The name was originally applied to a kind of lasting of which clergymen's gowns were made. It is now used to denote a variety of rich, satin-faced worsted cloth employed for women's dresses. The fibers are worsted. Prunella is dyed either in piece or yarn state and is hand finished.

Sacking. Plain solid color flannel in special shades for women's dressing sacks, also applied to a fabric made of hemp for grain sacks.

[Pg 99] **Sanglier.** A plain fabric of wiry worsted or mohair yarn, closely woven, with a rough finished surface. Sanglier is French for wild boar, the hairy, wiry cloth resembling the coat of the animal.

Sebastopol. A twill-faced cloth named from Sebastopol, the Russian fortified town captured by the English and French in 1855.

Serge. Under this name are classed a large number of fabrics of twill construction. In weight and texture a modern serge resembles flannel, except that it is twill woven and composed of fine yarn finished with a smoother surface. Serge comes from the Italian word *sergea*, meaning cloth of wool mixed with silk. Serges are woven of worsted, of silk, or of cotton yarn, and variously dyed, finished, and ornamented, as silk serge, serge suiting, storm serge, mohair serge, etc. Worsted serges of various kinds and degrees have been known since the twelfth century. Worsted serge appears to have come into general use as a material for men's wear in the sixteenth century. Modern serges vary but little from those made two centuries ago. They are dyed in a great variety of colors. On leaving the loom the cloth is washed and scoured with soap and water to remove the dirt and oil (if these remain the cloth will not take the dye properly). After dyeing, it is passed through a pair of metal rollers under pressure, which renders the surface more regular and even and of a better luster. This process accomplishes more than is required, for it produces a bloom on the surface which will show rain specks when in the garment, if it [Pg 100] is allowed to remain. This is ordinary serge. In order to make storm serge it is necessary to remove part of the bloom, and to accomplish this the cloth is steamed sufficiently to neutralize the effect of pressing. Steaming deadens the bloom and prevents the effects of rain showing on the cloth. The wearing qualities of serge are good, but it gets a shine easily. It is used for dress goods and suitings. Serge suiting used for men's clothing is a variety of light, wiry, worsted yarn woven with a flat twill, and dyed black or in shades of blue, fifty-four inches in width. Mohair serge is woven with a cotton warp and a mohair filling, thirty-two inches in width. This is dyed in a variety of colors and largely used as lining material for women's clothes, men's coats, and overcoats. Storm serge, designed to withstand exposure to stormy weather, is a coarse variety of worsted dress goods produced in a wide range of colors and qualities. The twill is wider, the texture stouter, and the surface rougher and cleaner than that of ordinary serge. Iridescent serge is a variety of worsted dress goods woven with warp and filling of different colors, causing a shimmering or iridescent effect. Cravenette serge is a fine twilled variety having a firm, closely woven texture, dyed black and in colors, and is used for women's gowns, men's summer suits, etc. Serge de Barry is a high-grade dress goods of fine texture, with fine twill, and wiry feel.

Shoddy is made from old woolen stockings or rags, shredded or picked by hand or machine, to render the yarn suitable for spinning a second time, or to give a [Pg 101] fiber that can be woven or felted with a wool or cotton warp. The name has come to mean cheap, make-believe.

Sicilian. Heavy weight cotton warp, mohair filled cloth. Sicilienne, the proper name, was made in the Island of Sicily as a heavy ribbed, all silk fabric.

Sultane. Twilled cloth of silk and wool; finished in the rough, not singed or sheared. The name is from Sultana, the first wife of the Sultan.

Tamise. Similar to etamine, with a very close mesh, made first of silk and wool. *Tamis* is French for sieve.

Tartans. Plaids of the Scottish clans worn by men in the Highlands of Scotland as a diagonal scarf, fastened on one shoulder and crossing the body. Each clan had a distinctive tartan or plaid. The name was adapted from the French *tiretaine,* a thin woolen checked cloth.

Thibet. Heavy, coarse weave of goat's hair, made by the Thibetans in Asia for men's wear.

Tricot. A heavy, compound fabric characterized by a line effect running warp way or filling way of the piece, usually produced with either woolen or worsted yarn. Tricot was originally a name given to fabrics made of woolen yarn or thread by hand knitting, and is the French word meaning knitting. The term was later applied to materials made on a knitting frame and now known as jersey cloth. Since 1840 the name tricot has been applied to finely woven woolen cloth, the weave of which is intended to imitate the face effect of a knitted fabric. The fabric is composed of woolen and worsted fibers, sometimes with cotton warp woven so as to hide [Pg 102] the cotton in finishing. The tricot line is similar to the rib line in a ribbed cloth except that it is not so pronounced. All tricots are constructed with two sets of warp thread and are characterized by a texture which, while dense, is singularly elastic, in this respect being somewhat similar to heavy jersey cloth. Tricots are commonly dyed in plain colors, and are finished clear so as to show the filling. When intended for trousers they are ornamented with small, neat patterns.

Tweed. A rough unfinished fabric of soft, open, and flexible texture, of wool or cotton and wool, usually of yarn of two or more shades; originally the product of the weavers on the bank of the river Tweed in Scotland. The face of the cloth presents an unfinished appearance rather than a sharp and clearly defined pattern.

Veiling includes light weight, usually plain weave fabrics of various constructions; generally made with singed or polished yarns. They are in solid colors. The use is designated by the name.

Venetian. Venetian cloth has a worsted or cotton warp and worsted filling; named from Venetia, a country around Venice. The warp yarns are firmly twisted, the twist being in the opposite direction to the twist in the filling yarn. Venetian is a trade term of wide application, in use since early times as a descriptive title for various fabrics, textures, and garments. One of the many varieties is a species of twill weaving in which the lines or twills are of a rounded form and arranged in a more or less upright position, hence a closely woven worsted cloth. The name is also applied to other fabrics, as a [Pg 103] twilled lining fabric woven with a cotton warp and a worsted filling known as Italian cloth. It is dyed in plain colors and is piece or yarn dyed for men. For women's wear it has light weight and plain colors with mixed effects and closely sheared nap. It is finished smooth so as to show the yarns prominently. Venetian cloth has not so much felting as broadcloth; it shows the weave more, but has the same lustrous finish.

Vigogne or **Vicuña.** A soft wool cloth of the cheviot order, with teasled face, made from the wool of the vicuña, a South American animal. Vigogne is the French name for the animal.

94

Vigoureux. A name applied to a plain or twill mixture, woven of undyed natural wool yarns. The French spinners found that the strongest yarns were those of the undyed wool. Sometimes two or more shades or tones are spun into one thread. The name is French for strong.

Voiles. Voiles are plain weave worsted fabrics made with hard twisted yarns. As clear a face as possible is secured in finishing, the cloth being singed or sheared closely if the yarns are not made comparatively free from loose fibers before being woven. Voiles are dyed in solid colors, and are used principally for dress goods.

Whipcord. Hard twisted worsted twills, either solid or mixed colors. The name is from the hard twisted lash of a whip.

Worsted Diagonals are characterized by prominent weave effects running diagonally across the cloth. The goods are usually of a solid color, and are given a finish [Pg 104] which brings the weave into prominence. Diagonals are used for suitings.

Unfinished worsted is a fabric woven with yarn with very little twist in it, and finished so as to make it appear covered with loose fibers, concealing the twill effect. After leaving the loom the cloth is placed in a fulling machine which condenses the fibers, thus increasing the density. It is then passed over hot presses after a slight shearing.

Finished Worsted is woven with yarn with a considerable twist, and finished in such a way as to show the construction of the cloth clearly. The finishing consists simply of scouring the cloth and not fulling it and then passing it through hot water baths between heavy rolls to remove all the soap. It is then sheared and pressed.

Zephyr. Light worsted yarn, also light weight cotton gingham. Zephyr is Greek for the light west wind.

Zibeline. A cloth manufactured with Merino lamb's wool for warp, and a light wool mixed with camel's hair for filling; or, worsted warp and camel's hair for filling; or either of the foregoing warps and a mixture of wool, camel's hair, and fine cashmere for filling. The long cashmere hair spreads over the surface. Used for ladies' tailor-made coats or suits, according to weight. The name is derived from the Latin word *sabellum*, meaning sable, and was applied originally to a variety of long-haired fur generally thought to be the same as sable. Zibeline has long hairs on its right side, some grades being almost like fur.

FOOTNOTE:

[15] Suggestions To Teachers. In connection with the study of fabrics the author has found it advisable to have the pupils insert in a blank book a sample of the fabric they are studying. In this way the pupil can examine both the filling (weft) and warp threads.

[Pg 105]

CHAPTER IX

COTTON

Cotton. Cotton is the most important vegetable fiber used in spinning. The cotton fiber is a soft, downy substance which grows around the cotton seed. When examined under the microscope it appears as a long twisted cell. Owing to the fact that the cotton-plant yields so readily to the varying conditions of soil and climate, there is a large variety of cottons, each having some peculiarity which is considered enough to place it in a distinct class. An idea of the number of species of the cotton-plant can be obtained from the fact that the United States Department of Agriculture has recorded about one hundred and thirty varieties. The most important varieties are: *Gossypium herbaceum, G. arboreum, G. hirsutum, G. barbadense,* and *G. peruvianum.* The botanical name of a plant is divided into two parts: first the family name, followed by the species name.

The *Gossypium herbaceum* grows from four to six feet in height and bears a yellow flower. The seeds are covered with a short gray down. The fiber it bears is classed as short. It is found in Egypt, Asia Minor, Arabia, India, and China. The short-stapled variety of Egyptian cotton is from this species.

[Pg 106] The *G. arboreum* when full grown attains a height from fifteen to twenty feet. The seed is covered with a greenish fur and is enveloped in a fine, silky down, yellowish white in color. It is found in Egypt, Arabia, and China.

The *G. hirsutum* is a shrubby plant, its maximum height being about six feet. The young pods are hairy, and the seeds are numerous and covered with a firmly adhering green down. It is probable that this is the original of the green-seeded cotton which is now cultivated so extensively in the Southern States of America, and which forms the bulk of the supply from that source.

The *G. peruvianum* is similar to the *G. barbadense.* The Brazilian and Peru cottons are from this species.

The *G. barbadense* grows from six to fifteen feet high; its flowers are yellow and its seeds black and smooth, being quite destitute of the hair that distinguishes other members of the species. It is a native of Barbadoes or has been cultivated there for a long time. Cottons of the finest texture belong to this species—Sea Island and Florida cottons—from which our finest yarns are spun, and it is used chiefly in the manufacture of fine lace. The long-stapled Egyptian and several other varieties are said to be from this stock.

Cotton Growing Countries. The most suitable situation for growing cotton is between 35 degrees north and 40 degrees south of the equator. The chief cotton growing countries of the world in order of importance are: United States, India, Egypt, and Brazil. Cotton [Pg 107] is also grown in the following countries, but in no quantity or quality comparable with the four named above—West Indies, west coast of Africa, Asia Minor, China, and Queensland.

The best soil for growing cotton is a light loam or sandy soil, which receives and retains the heat, and at the same time preserves a good supply of moisture. Cold, damp days are not suitable for its growth, while deep rich soils develop too much leaf and stalk. The best climate for the cultivation of cotton is where frost and snow are of short duration, dews are heavy, and the sun bright, warm, and regular. New soils generally produce the best cotton. The character of the cotton fiber is dependent upon three things, the species of the plant, the nature of the soil, and the locality in which it is grown.

Rough Peruvian. The nature of this cotton is harsh and wiry and resembles wool so nearly that it is almost exclusively used to mix with woolen fabrics. The staple is rough and generally strong, and is of a springy tendency, *i.e.*, it does not lie close like American.

East Indian. India depends upon the monsoon for its moisture, and the success or failure of the crop is decided by that phenomenon of nature. Indian cottons as a rule are coarser and shorter than American cottons. The land is prepared before the breaking of the monsoon, and the planting begins after it. There is not the same care bestowed upon the cultivation of the Indian cotton, nor are such improved methods practised as in America. The ancient routine of past generations [Pg 108] still persists, and as a consequence the yield per acre is less than one-half that of America. Moreover the acreage planted is only about two-thirds that of America. The better growths of East Indian cotton were once largely used in this country for filling, owing to their good color and cleanliness; but of late years the consumption has steadily decreased, owing chiefly to the increased takings by the Indian mills, also to the exports to China and Japan, and to the preference shown by English spinners for American cotton.

Egyptian Cotton. Egyptian cotton, on account of its long staple and silky gloss, is imported in considerable quantities. Egyptian is largely used in the manufacture of hosiery, and also for mixing with worsted yarn. Owing to its gloss it is used for mixing with silk, and on account of its strength it is made into the finer sewing threads. Egyptian cotton is sometimes so charged with grease that it has a greasy smell; and to make it workable it is necessary to sprinkle it with whitening. It has been observed that velvets woven (or piled) with Egyptian filling do not finish as well as when picked with yarns made from American cotton, the reason for this being that the greasy nature of the Egyptian cotton fiber often varies in strength, causing different shades in the finished goods. This greasy nature is said to be due to two things: (1) the fertility of the soil; (2) the extent to which the cell walls of the fibers are developed.

In addition to cotton, other crops are grown in Egypt—rice, sugar, beans, barley, onions, etc.—and the [Pg 109] acreage devoted to cotton is regulated to some extent by the prospects as to which crops are likely to pay best. It is calculated that not more than one-third of the area is usually devoted to cotton.

Sea Island Cotton. This is the finest growth of cotton, and it commands the highest price. The staple, which is long and silky, varies in length from one and a half to two and a half inches. It is used for making fine muslins, laces, spool cotton, and other fabrics, and is also largely mixed with silk. It is said that this cotton was first introduced into America in 1786 from the Bahama Islands, whither it had been brought from the West Indies. It was first cultivated in Georgia, where it was found that the small islands running along the coast were best adapted for its growth, hence the name "Sea Island." It was also grown on the uplands of Georgia, but although remaining good, the quality deteriorated. Counts as high as four hundred are occasionally spun in Sea Island cotton.

Other Varieties. Cotton grown in the Southern States under widely varying conditions of the soil, climate, and care in cultivation, naturally varies in length, strength, and other qualities of staple. Cotton known as "Uplands" or "Boweds" varies in length from three-fourths to one and one-sixteenth inches and is used for filling; this is grown in North and South Carolina, Georgia, Florida, Alabama, and Tennessee. Cotton used for twist is grown in Texas, Louisiana, Mississippi, and Arkansas, and the length of the staple varies from one to one and three-sixteenths inches. In [Pg 110] the swampy and bottom lands in some of the states (notably Alabama, Louisiana, Mississippi, and Arkansas), cotton is grown with staple ranging from one and one-eighth to one and one-fourth inches. In addition to these, there are especially long stapled growths, known as "Extras," "Allen Seed," and "Peelers," which measure one and three-eighths to one and five-eighths inches. Of late there has been an extensive demand for long-stapled American cotton (one and three-sixteenths to one and one-half inches), owing to the development of fine spinning.

Cotton Raising. Cotton is planted with a machine, which puts it under the ground about one and one-half to two inches. It is not planted as corn is, that is, dropped so far apart, but is planted in a continuous stream. After the cotton comes up out of the ground, when it is about three inches high, it is hoed by ordinary labor with a hoe, and is cut out or, rather, thinned. This is called "chopping out" and is for the purpose of removing the inferior or weak plants until only one strong plant is left. The distance between the plants depends on the nature of the plant, frequently about twelve inches being left between them.

The American Crop. The first step taken is the preparation of the ground for planting. This begins in the southern part of Texas as early as the middle of January, in Florida about the third week; in Alabama, Georgia, Mississippi, and Louisiana, about the beginning of February; in Arkansas, Tennessee, and South Carolina from about the middle of February to the [Pg 111] beginning of March. Actual planting begins according to latitude, principally from the middle of March to the middle of April, and ends in the first half of May. These dates, however, are dependent upon the state of the weather. When the weather is unusually wet the start is late. The plant suffers from the rank growth of grass and weeds, and extra labor is required to keep the fields clean. In abnormally hot weather, especially after rains, the plant sheds its leaves, thus exposing the bolls, which fall off, whereupon replanting becomes necessary. In addition to injuries by the weather the cotton-plant is subject to depredations by insects. Of late years the greatest pest has been the Mexican boll weevil.

The cotton-plant blooms ten or eleven weeks after planting. An early bloom is taken as a sign of good crops. When the crop is an early one, picking may commence in the districts in which it ripens first in the latter half of July; but the usual date is the beginning of August, following on in the various districts in succession until the early part of September. The plant goes on fruiting as long as the weather is mild and open. It finishes in the early regions about the beginning of December, the others following through December and closing in the later regions about the middle of January. Frosts play an important part in the ultimate yield. An early killing frost over the entire belt would curtail the size of the crop by 500,000 bales in a season, as was the case in 1909 when about 32,000,000 acres were planted. Light frosts and late frosts do little harm to the cotton-plant; in fact it is contended that the late [Pg 112] frosts do much good under certain conditions of the crop, by opening the bolls that otherwise would not open, and thus adding to the quantity of the late pickings. The effect of frost upon the lint so picked is to produce tinged and stained cotton. Early killing frosts occur in some seasons in the early part of November, when much of the yield may be curtailed. When killing frosts occur late in the season, when the fruiting is practically over, it has little or no effect upon the yield except as regards the color.

The ripening of the crop proceeds in three stages, the bolls nearest the ground maturing first, then those around the middle of the plant, and lastly the top crop. Pods half ripe are often forced open and the fiber sent on with good cotton. East Indian is more highly charged with unripe cotton than American. The work of picking is not heavy, but becomes tedious from its sameness. Each hand as he goes to the field is supplied with a large basket and a bag. The basket is left at the head of the cotton row, the bag being suspended from the picker's shoulder by a strap, and used to hold the cotton as it is plucked from the boll. When the bag is full it is emptied into the basket, and this routine continued throughout the day. Each hand picks from 140 to 180 pounds of cotton per day. The average yield in the South varies from 500 to 600 pounds per acre. Every boll of cotton contains seeds resembling unground coffee; when these have been removed by the gin, there remains about one-third the weight of the boll in clean cotton.

[Pg 113]

PICKING COTTON

[Pg 114] **Ginning.** The next operation to which cotton is subjected is that of ginning, or separating the seeds from the fiber. This work was formerly accomplished by hand, and so great was the quantity of seeds that frequently an entire day was occupied by a workman in separating them from one pound of cotton. At the present day the devices for separating the lint from the seed are of two classes: roller gins and saw gins. The former device is the more ancient, having been used from the earliest times by the Hindoos. In its simplest form it consists of two rollers made of metal or hard wood, fixed in rude frames, through which the cotton is drawn and the seeds forced out in the process. An improved form of the roller gin is at present used for cleaning the long-staple Sea Island cotton. The saw gin, which works on an entirely different principle, is the machine which, with its improvements and modifications, has separated the seed from fiber almost exclusively for one hundred years of American cotton growing. In this machine the seed cotton is fed into a box, one side of which is formed of a grating of metal strips set close together, leaving a narrow opening from one-eighth to a quarter of an inch wide. Into these openings a row or "gang" of thin circular saws project mounted upon a revolving mandrel. The long, protruding teeth of the saws, whirling rapidly, catch the fibers, and pull them away from the seeds. The latter, being too large to pass through the openings of the grating, roll downward and out of the machine. The lint, removed from the row of saws by a revolving [Pg 115] brush, passes between rollers and is delivered from the machine in the form of a lap or bat.

This machine is responsible for much of the "nep" (or knots) found in American cotton, which is caused when the machine is overcharged. The Whitney gin will turn through more cotton than any other type of machine, and will clean from 200 to 300 pounds per hour. When the machine is running at high speed the tendency is to string and knot the cotton.

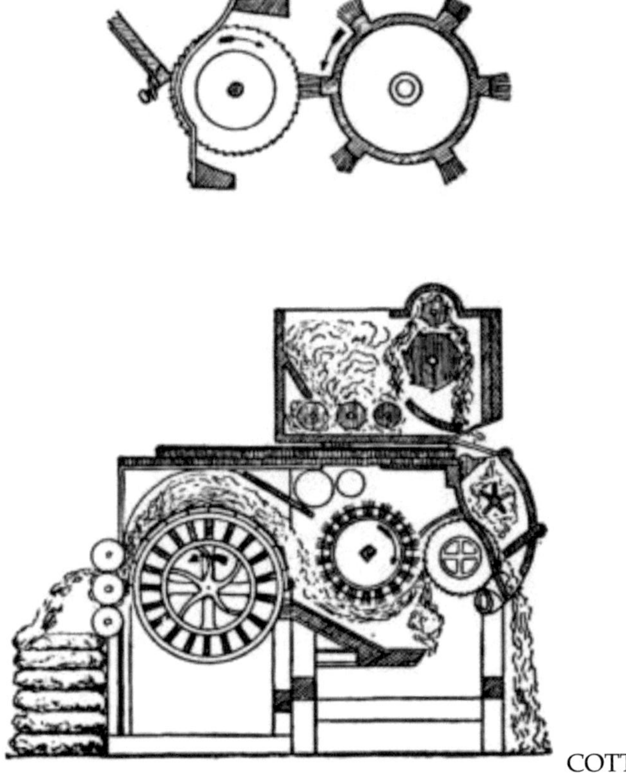

COTTON GIN

The upper figure shows Whitney's invention. The lower figure
shows a later form.

The working of the ordinary gin is as follows: The wagon loaded
with cotton is driven under a galvanized spout called the sucker,
through which there is a suction of air which draws the cotton into
the gins. In each of the gins there are seventy circular saws revolving on one shaft. These saws are about one inch apart, and the teeth
go through the gin breast, much as if one were to put the teeth of
one comb into the teeth of another comb. This process takes the lint
cotton off the seed, and by the use of brushes the cotton goes into
the lint flute, into the condenser, and into the box, where it is revolved and made into a bale. While the lint is going through this
process, the seeds, being heavier and [Pg 116] smaller, draw to the

104

bottom of the gins, fall into an auger which is operated by a belt, and then are dropped into a conveyor and carried to the seed pile or houses. The lint goes in one direction and the seed in another.

When the seed is taken from the cotton at the gin, it is covered with a lint of cotton. In order to remove this the seeds are put through a delinter, which takes off the small, short fiber from the seeds, leaving them clean. This seed is then put through a huller which takes off the outside hull or thick skin. The kernel is then put through a hydraulic press, which squeezes the cotton-seed oil from it and leaves the "meal." Cotton-seed oil is used for many purposes, such as making olive oil, butter or oleomargarine, lard, etc. Of late an experiment has been made with the meal for use in the place of flour, and has been pronounced a success. Seed crushing has now become an important industry, with the cotton crop each year amounting to between 12,000,000 and 13,000,000 bales of 450 pounds each.

The Cotton Gin. The cotton gin was invented in 1792 by Eli Whitney, a citizen of Georgia, but a native of Massachusetts. The importance of this invention to the cotton industry of the world cannot be overestimated. It was the one thing needed to insure a sufficient supply of raw material to meet the requirements of newly invented machinery for spinning and weaving. The result of Whitney's invention was the rapid extension of the culture of cotton in the United [Pg 117] States, and its permanent establishment as one of the leading staples of the country.

Cotton Bales. After the cotton is ginned and baled it is shipped to the mill. The standard size of a cotton bale in the United States is 54 × 27 × 27 inches, and contains nearly 500 pounds. To produce this bale over 1,600 pounds of seed cotton are required. The bales are wrapped in jute bagging and strapped with sheet-iron bands, this covering adding about twenty-five pounds to the weight of the bale.

The Bessonette cylindrical bale is turned out by a self-feeding press, which receives the lap of lint from the gin between two heavy rollers. The fiber is rolled upon a long wooden spool so tightly as to press out nearly all the air, and forms a package of uniform shape and size throughout, having a diameter of fourteen to sixteen inches. The bales are covered with cotton cloth, held in place by small wire hoops. It is claimed that the cotton is rolled so tightly by this process that the bales are practically fireproof and waterproof.

Egyptian bales are compressed into a shape similar to the American bale, but the average weight is over 700 pounds.

The Indian bales, which are more closely compressed than the American, usually weigh 400 pounds.

Cotton is purchased by the mill authorities in the shape of a bale. The method is to purchase from cotton brokers, samples being furnished to the buyer from which to make selection.

[Pg 118]

STOREHOUSES

[Pg 119] The commercial value of cotton is determined by its length, fineness, strength, pliability, smoothness, regularity, color, and cleanliness. As a rule, the cotton that is the longest is also the finest, but by no means the strongest. Thus, Sea Island cotton has the longest staple with the least diameter, and Hinganghat (an Indian cotton) is much inferior to it in both respects. The strength of the latter, however, is 50 per cent greater than the strength of Sea Island cotton. In every other respect Sea Island cotton is in advance over Hinganghat cotton. It is the most valuable, especially for the production of fine yarns.

106

The most regular cotton is Orleans, in which the length of the staple varies only a small fraction of an inch. In consequence of this there is less loss in working Orleans than is the case with the other cottons, owing to the fact that their fibers vary in length.

The Leading Growths of Cotton. In order to purchase the raw material of the cotton manufacture, to arrange the "mixing" or have much to do with the raw material in any other capacity, one should know as much as possible of its characteristics; for ignorance may cause much trouble and no little loss to those who have to spin the cotton. Each crop differs from the previous one to a greater or less degree, as it depends entirely upon the weather. Thus, in a very dry season there is a "droughty crop" which, while it may be (and generally is) clean and well up in class, will be weak, short, and of irregular fiber. In order to obtain the desired length and strength of staple the buyer will [Pg 120] have to pay a relatively higher price than in what may be termed a normal season.

FANCY COTTON LOOM

Again, in a crop that is poor in class, a defect that may have been caused by too much rain in the early or middle stages of its growth, or by unfavorable weather for the production of cotton of good grade, the staple will probably be all that could be desired, leafy and small, but the buyer will have to pay more to obtain his usual grade, especially if he requires it for good [Pg 121] filling. Then there are seasons when the crop turns out fairly well in class and staple, but the cotton is wasty, dirty, or abnormally leafy; and in this case the buyer has to exercise great care and judgment in calculating the extra loss that will ensue.

The terms of purchase of cotton include an allowance of 4 per cent for tares. That is, a bale of cotton weighing 400 pounds would be paid for as 384 pounds, or should the buyer have reason to believe that the tares are unusually heavy, he has the option of claiming the actual tare. This is ascertained by stripping ten bales and weighing the covering and the hoops, which means considerable work, and although it is at the option of the buyer, it is an exception rather than the rule.

As a result of these causes we find cotton divided into the following grades:

Full Grades of Cotton. Egyptian cotton is graded as follows: extra fine, fine, good, fully good fair, good fair, fair, middling fair, middling.

Indian cotton is graded as follows: superfine, fine, fully good, good, fully good fair, good fair, fully fair.

Brazilian cotton may be classed: fine, good, good fair, fair, middling fair, middling.

American cotton has seven grades: fair, middling fair, good middling, middling, low middling, good ordinary, and ordinary.

In addition to the full grades there are half and quarter grades. The American cottons are graded as follows:

[Pg 122]

Full Grades.	Half Grades.	Quarter Grades.
Fair,	Strict middling fair,	Barely fair,
Middling fair,	Strict good middling,	Fully middling fair,
Good middling,	Strict middling,	Barely middling fair,
Middling,	Strict low middling,	Fully good middling,
Low middling,	Strict good ordinary,	Barely good middling,
Good ordinary,	Strict ordinary.	Fully middling,
Ordinary.		Barely middling,
		Fully low middling,
		Barely low middling,
		Fully good ordinary,
		Barely good ordinary.

The following are a few of the leading varieties of cotton, with the numbers of yarn they will make:

Cotton.	Length.		Warp.	Filling.
Sea Island (selected)	1¾	to 2¼	up to 200	250 to 300
Sea Island (ordinary)	1¾	to 2	150	220
Florida Sea Island	1¾	to 2	150	220
Georgia	1½	to 17/8	120	180
Egyptian	1¼	to 1½	70	120
Peeler	1¼	to 13/8	50	70
Orleans or Gulf	11/16	to 1¼	40	60
Upland	1	to 11/8	30	45
Texas	7/8	to 11/16	25	35

During the last few years considerable discussion has taken place among mill men, both in this country and abroad, bearing upon the subject of moisture contained in baled cotton. Of course the natural moisture in the cotton fiber varies, as might be expected, from year to year, according to the character of the season during the picking. The standard of moisture is based upon what is known as regain, that is, if 100 parts of absolutely dry cotton are exposed to the air, they will [Pg 123] absorb about 8½ per cent of moisture, although a much higher per cent is sometimes found.

In some of the small Southern mills located in the cotton raising section, the cotton is delivered by team direct from the gin, without going through the compress. In this way they save the greater part of transportation expense. They also save in the strength of the cotton fiber itself, since the process of compression injures the fiber. They get better cotton, being nearer the source of supply and having better opportunities for selection.

When the cotton arrives in the shape of a bale, it is necessary to cut ties and loosen up the cotton before use. This may be done in two ways. One method being to pull the bale apart by hand, and the other to pass it through a bale breaker or similar machine, which loosens up the cotton by means of beaters. It now starts on a continuous journey through successive machines until it is made into yarn. The yarn is made into a warp, and the warp interlaced with the filling yarn to make cloth, and the cloth finished for the market.

Not every country is adapted for making cotton yarn, for certain conditions are necessary to manufacture good yarn. If the atmosphere is too warm or too dry, the fibers will become brittle and will not twist well; if too wet they collapse and stick. Lancashire County, England, seems to have been fitted by nature for cotton spinning. It has just the right climate, a moist temperature, and copious water supply. There are hills on the east of the valley, forming a water shed, and the [Pg 124] town lies in a basin covered with a bed of stiff clay, that holds the water, allowing it to evaporate just fast enough to keep the air in the moist condition needed to fit the fibers for weaving. Countries that have not these conditions are obliged to produce them by artificial means—humidifying, etc.

PICKER ROOM
1. Hopper where the cotton from the bale is fed into Picker.
2. "Lap" showing how the cotton is prepared for the card.
3. Picker Machine (complete).

[Pg 125]

CHAPTER X

MANUFACTURE OF COTTON YARN

PICKER ROOM—SHOWING END VIEW OF PICKER

1. Lap of Cotton.

Picker Room. The first step in the conversion of the bale of cotton into yarn consists in giving the cotton fibers a thorough cleaning. This is accomplished by feeding the cotton to a series of picker machines called in order, bale breaker, cotton opener and automatic feeder, breaker picker, intermediate picker, and [Pg 126] finisher picker. These machines pull to shreds the matted locks and wads of cotton (as we find them in the bale), beat out the dirt, stones, and seeds, and finally leave the cotton in the form of batting upon the cylinders; this batting passes from one machine to another until it issues from the finisher picker as a downy roll or lap.

(Sometimes the bale breaker is not used in the mill.)

CARD ROOM

1. Roving Can — receptacle to hold the sliver. After it is filled it is transferred to either ribbon lap machine or drawing frame.
2. Cylinder of the card. The cotton is on this cylinder in the form of a web.

Carding Machine. When the lap of cotton leaves the picker it goes to the carding machine, where it is combed into parallel fibers by means of a revolving cylinder covered with wire teeth called card clothing. [Pg 127] As the cotton is fed to the card in the form of a sheet or lap from the picker, it is supposed to have been freed from a considerable quantity of sand, seed, etc., but there still remain nep, fine leaf, and short fibers, which are removed during carding.

On leaving the card cylinder the lap has become a gossamer-like web thirty-nine inches broad. This web next passes through small "eyes," which condense it into a narrow band about an inch in width, known as card sliver.

When a lap is delivered from the finisher picker, it should weigh a given number of ounces per yard. The method of ascertaining the weight is to make each lap a standard number of yards in length

114

and weigh each lap. The machine can be regulated so as to give the desired weight per yard.

Combing. When an extremely fine and strong yarn is required, in addition to carding, the fibers are also subjected to the process of "combing." This may be said to be merely a continuation of the carding process to a more perfect degree. The chief object is to extract all fibers below a certain required length, and cast them aside as "waste." This is done in order to secure the very best fibers calculated to give the strongest and best results in the spun yarn.

The process of combing follows carding. The card delivers the cotton in the form of a sliver or strand, while the combing machine requires the fibers to be delivered to it in the form of sheets, nine to twelve inches wide. This is done by taking a number of card [Pg 128] slivers and forming a lap of them by passing the sliver through a sliver lap machine. The laps are passed through the comber. This machine consists essentially of a series of rollers, nippers, and rows of metal teeth. By the action of these, the short fibers are separated and combed out, and the long ones arranged in parallel order in the form of a thin, silky strand, in which condition it is sent to the drawing frames to be drawn out. Of course it must be understood that a combing machine is used by only a small percentage of cotton spinners. For ordinary purposes a sufficiently good quality can be made without a comber. As there is from 15 to 35 per cent waste to this operation it may be readily seen that it is costly, and limited entirely to the production of the very best and finest yarns, such as those intended for sewing or machine thread, fine hosiery, lace curtains, underwear, imitation silks, and fine grades of white goods. There are combing machines that comb short staple cotton.

Drawing. The cans containing the slivers are taken from the card or combing machine (as the case may be) to the drawing frame. The object of this machine is mainly to equalize the slivers, combining a number of them together so as to distribute the fibers uniformly. The condition of the fibers on leaving the card or comb is such that a slight pull will lay them perfectly straight or parallel, and this pull is given by the drawing frame rollers. Of course the fibers coming from the comb are parallel, but it is necessary to alternate them by the drawing. The drawing frame is a machine consisting of a number of sets of rollers, the front roller having a greater speed than the rear ones.

[Pg 129]

COTTON COMB ROOM

1. The cotton in the form of a "lap" ready to pass through the comb.

[Pg 130] The slivers, which are as nearly as possible the same weight per yard, are combined together in the drawing and emerge from the pair of front rollers as one sliver weighing the same number of grains per yard as a single sliver fed up at the back. This process is repeated two or three times, according to requirements, the material then being referred to as having passed through so many "heads" of drawing. It is not unusual to pass Indian and American cotton through three deliveries.

The object of all the processes thus far described has been that of cleaning (in the picker), arranging the fibers in a parallel position to each other, making uniform, and drawing out the stock. In every case the stock delivered from a machine is lighter than when fed into it, and contains just twist enough to hold it together and prevent its being stretched or strained when unwound from the bobbin, and fed into the next machine. The minimum amount of twist in roving is desirable for the reason that it permits the stock to be

116

drawn out more easily and uniformly, the little twist that is put in the roving by the slubber being practically eliminated when it is passed through the rolls of the intermediate. The same applies in the case of the roving passing from the roving to the spinning frame.

Fly Frames. The process in the manufacture of yarn after the cotton has passed through the drawing frame consists of further attenuation of the sliver, but as the cotton sliver has been drawn out as much as is [Pg 131] possible without breakage, a small amount of "twist" is introduced to allow of the continued drawing out of the sliver.

From the drawing frame, the drawing passes through two, three, or four fly frames, according to the number of yarn to be made. All these machines are identical in principle and construction, and differ only in the size of some of the working parts. They are the slubber, intermediate, roving,—and fine or jack frame-fine, and the function of each is to draw and twist.

ROVING DEPARTMENT

1. Slubber machine, showing sliver of cotton passing through the rolls and then given a twist while it is wound on the bobbin.

Intermediate Frame. The function of the intermediate frame is to receive the slightly twisted rove from the slubber and add thereto a little more twist [Pg 132] and draft. The rove is taken from two bobbins to one spindle in the machine, an arrangement which tends to insure strength and uniformity. The principle of the machine is in other respects the same as that of the slubbing frame.

Roving Frame. The function of the roving frame is to receive the twisted rove from the intermediate and add more twist and draft, thereby further attenuating the rove. As in the intermediate frame the rove is generally taken from two bobbins for one spindle.

Fine or Jack Frame. This machine is used when fine yarns have to be made. It is built on the same principle as the preceding frames, the only difference being that a finer rove is made from which finer numbers of yarn can be spun. As in the slubber, intermediate, and roving frames, the rove is taken from two bobbins for one spindle.

Spinning. In the manufacture of single ply yarn the final process is that of spinning, which consists in drawing out the cotton roving to the required size, and giving it the proper amount of twist necessary to make the yarn of the required strength. While the spinning frame is built on entirely different principles from the roving, intermediate, or slubber frame, the object of each machine is the same as that of the spinning frame. The principal point of difference is the amount of twist imparted to the cotton roving.

[Pg 133]

ROVING ROOM

1. A drawing frame showing the sliver of cotton passing through the machine.

2. A slubber showing the sliver passing through and wound on bobbins.

3. Roving machine showing the cotton passing from one bobbin through the roller to another.

[Pg 134] The objects of the spinning process are:

1. Completion of the drawing out of the cotton roving to the required size.

2. Insertion of the proper amount of twist to give the thread produced strength.

Excessive speed causes defects in the yarn and undue wear and tear on the machine.

There are two methods of spinning: ring spinning and mule spinning. The mule spinning is the older form. There are but few mule frames in operation in this country.

Mule Spinning. The function of mule spinning is to spin on the bare spindle, or upon the short paper tubes, when such are required to form a base for the cop bottom. The mule will spin any counts of yarn required, and is especially adapted for yarn in which elasticity and "cover" are essentials. Hosiery yarns are produced on the ordinary cotton mule and are very soft spun.

The bobbins of roving are placed in a creel at the back of the machine, the stands of roving being passed through the rolls and drawn out in the same manner as at the roving frame. The spindles are mounted on a carriage which moves backward and forward in its relation to the rolls, the distance roved being about five feet. When the spindles are moving away from the frame the stock is being delivered by the rolls, the speed at which the spindles move away from the rolls being just enough to keep the ends at a slight tension. The twist is put in the yarn at the same time.

[Pg 135]

SPINNING ROOM. COTTON DEPARTMENT
1. Humidifier—an apparatus to give off moisture.
2. Spinning frames—showing the cotton as it comes from the roving frame and passes through the spinning frame.

[Pg 136] When the spindles reach their greatest distance from the rolls, the latter are automatically stopped and the direction of the motion of the spindle carriage reversed. The yarn is wound on the spindle while the carriage is being moved back toward the rolls, the motion of the rolls being stopped in the meanwhile, the spindles revolving only fast enough to wind up the thread that has been spun during the outward move of the carriage.

The mule is a much more complicated machine than the ring frame, its floor space is much greater, and more skilled help is required for its operation. Under ordinary conditions it is not practical

120

to spin finer yarn than No. 60s on a ring, while as high as No. 500s is said to have been spun on a mule. The same number of yarn can be spun on a mule with less twist than on the ring. This is important in hosiery yarn.

Ring spinning is used for coarse numbers, and has greater production and requires less labor than mule spinning. Ring-spinning yarn is used for warp purposes.

Ring Spinning. The function of ring spinning is to draw out the rove and spin it into yarn on a continuous system. The yarn made is spun upon bobbins.

The ring spinning differs from mule spinning in having the carriage replaced by a ring, from which the machine takes its name. The ring is from one and one-half to three inches in diameter, grooved inside and out, and is connected with a flat steel wire shaped like the letter D, called the "traveller." Its office is to constitute a drag upon the yarn, by means of which the latter is wound upon a bobbin. Its size and weight depend on the counts of yarns to be spun; coarse yarns demand the largest ring and heaviest traveller.

[Pg 137]

MULE ROOM
1. Stocks from Finisher ready to be spun upon the mule.
2. Front Rollers with weight levers.

3. Clearers for Front Rollers.
4. Faller shafts containing Sickles.
5. Spindles.
6. Spun Yarn wound on cops.
7. Covers for carriage.
8. Body of carriage.
9. Drawing out band.
10. One of the wheels for the carriage.

[Pg 138]

CHAPTER XI

THREAD AND COTTON FINISHING

Thread. In general a twisted strand of cotton, flax, wool, silk, etc., spun out to considerable length, is called thread. In a specific sense, thread is a compound cord consisting of two or more yarns firmly united by twisting. Thread is used in some kinds of weaving, but its principal use is for sewing, for which purpose it is composed of either silk, cotton, or flax. Thread made of silk is technically known as sewing silk; that made of flax is known as linen thread; while cotton thread intended for sewing is commonly called spool cotton. These distinctions, while generally observed by trade, are not always maintained by the public.

The spool cotton of to-day is of a different grade from that made before the sewing machine came into general use. The early thread was but three cord, and contained such a large number of knots, thin places, etc., that it could not be worked satisfactorily on the machines, so manufacturers were called upon to produce a thread that would be of the same thickness in every twist. This was effected by making the thread of six cords instead of three, thereby producing a smoother and more uniform strand.

[Pg 139] **Manufacturing Processes.** The raw cotton for the manufacture of thread must be of long staple. If the fiber is short the thread made of it will be weak, and hence unsuited for the purposes required of it. Ordinary cotton is not adapted to the manufacture of the better grades of spool cotton on account of the shortness of its fiber. Egyptian and Sea Island cotton are used because they have a much longer fiber and are softer in texture. The raw cotton comes to the factory packed in great bales, and is usually stored away for some months before it is used. The first step in the conversion of the bale of cotton into thread consists in giving the fiber a thorough cleaning. This is accomplished by feeding it to a series of pickers which pull the matted locks and wads to shreds, beat out the dirt and seeds, and roll the cotton in the form of batting upon cylinders until it issues from the finisher lap machines as a downy roll or lap.

The lap of cotton then goes to the carding rooms, where it is combed into parallel fibers by means of a revolving cylinder covered with fine wire teeth, sometimes 90,000 of them to the square foot. On leaving the carding machines the lap has become a gossamer-like web thirty-nine inches broad. This web is next passed through a small "eye" which condenses it into a narrow band about an inch in width, known as the sliver. By this time the fiber has been so drawn out that one yard of the original lap has become 360 yards of the sliver. The sliver now looks almost perfect, but if it were spun it would not make good thread. It is necessary [Pg 140] to lay every fiber as nearly parallel as possible, so that there will be an equal number of fibers in the strand per inch. Besides this, the remaining dirt and short fibers must be removed and the knots and kinks in the fibers straightened out. To accomplish these objects the cotton must be "combed." First, the slivers are passed through several sets of rollers, each set moving faster than the preceding, so that the strands are drawn out fine and thin. In this condition the cotton passes to a doubling frame, and from thence to the lapping frame, a device combining six laps into one and drawing the whole out into one fine, delicate, ropy lap.

WARP ROOM

1. Beam on which the warp is wound.
2. Warp.
3. Creel.
4. Spools in the creel.

[Pg 141] The comber now takes the lap and combs out all the impurities and short fibers, at a sacrifice of about one-fifth of the material; next, it combines six of these fluffy combed rolls of fiber into one. A number of these rolls are then drawn out by another machine twelve times as long as they were before and twisted together on a slubbing frame. This last drawing reduces the roll to about the thickness of zephyr yarn. After being further doubled and twisted, the yarn, or roll, is ready for the mule spinner, which accomplishes by means of hundreds of spindles and wheels what the housewife once did with her spinning wheel. The mule, however, does the work of more than 1,000 hand spinners and takes up much less space. On this machine 900 spindles take the yarn from 1,800 bobbins, and by means of accelerating rollers and a carriage draw out and twist it to the proper fineness for the size of thread wanted. Having passed through the complex processes of cleansing, combing, drawing, and spinning, the cotton is now in the form of yarn of various sizes, and the real work of thread making, which is a distinct art from yarn making, begins.

The thread-making process is briefly as follows: The yarn is doubled and twisted; then three of such yarns are twisted together, which give the six-fold combination for six-cord thread. For a three-cord thread three yarns are twisted together. After the twisting is completed the thread is reeled into skeins having a continuous length of 4,000 to 12,000 yards, according to the size, and is then sent to the examining [Pg 142] department where it is rigidly inspected. Every strand is looked over, and any found to be defective are laid aside, so that when the thread is put on the market it shall be as perfect as care and skill can make it.

At this stage of the work the skeins of thread are of the pale cream color common to all unbleached cotton goods, and are technically known as "in the gray." They therefore have to be bleached pure white or dyed in fast colors. The skeins, whether intended for white or colored thread, are first placed in large, steam-tight iron tanks and boiled. Here the thread remains subjected to a furious boiling for six or seven hours; when removed it is perfectly clean, but still retains the brownish gray color of unbleached cotton. It then goes into a bath of chloride of lime and is bleached as white as snow. The skeins are next drawn through an acid solution to neutralize the chloride. Another boiling, another bleaching, a bath of soapsuds, and the final rinsing, complete the cleansing and whitening process. Those skeins intended for colored threads are taken to the dyeing room and placed in tanks filled with suitably prepared dyeing solutions.

[Pg 143]

TWISTING ROOM
1. Humidifier.
2. Twister machine.
3. Boxes containing spools of cotton, ready to be put in creel and form warp.

[Pg 144] From the bleaching and dyeing departments the skeins of thread go back to the mill to be wound on the bobbins, and from the bobbins finally on the small wooden spools. The automatic winding machines can be regulated to wind any given number of yards. The small spools are fastened on pivots, the thread from the bobbins fastened on the spools, and the machines set in motion. At the required number of yards the spools stop revolving. The ordinary spool of cotton thread contains 200 yards, and when this has been wound on, the thread is cut with a knife by an attendant, who also cuts the little nick in the rim of the spool and fastens therein the end of the thread. Thread mills commonly print their own labels, and these are affixed to the spools by special machinery with re-markable rapidity. From the labeling machine the spools go to an inspector, who examines each one for imperfections, and any that are found faulty are discarded. When packed in pasteboard boxes or in cabinets the thread is ready for market.

Thread Numbers. Spool cotton for ordinary use is made in sizes ranging from No. 8 coarse to No. 200 fine. In cotton yarn number-ing, the fineness of the spun strand is denoted by the number of hanks, each containing 840 yards, which are required to weigh one pound, as illustrated in the following table:

When 1	hank of cotton yarn	(840 yds.)	weigh	1 lb.	it is	No. 1
"	10 " " " "	(8,400 yds.)	"	"	" "	10
"	16 " " " "	(13,440 yds.)	"	"	" "	16
"	30 " " " "	(25,200 yds.)	"	"	" "	30
"	50 " " " "	(42,000 yds.)	"	"	" "	50
"	100 " " " "	(84,000 yds.)	"	"	" "	100

The early manufactured thread was three cord, and took its num-ber from the size of the yarn from which it was made. No. 60 yarn made No. 60 thread, though in point of fact the actual caliber of No. 60 thread would equal No. 20 yarn, being three No. 60 strands [Pg 145] combined together. When the sewing machine came into the market as the great consumer of thread, spool cotton had to be made a smoother and more even product than had previously been necessary for hand needles. This was accomplished by using six strands instead of three, the yarns being twice as fine. As thread numbers were already established, they were not altered for the new article, and consequently at the present time No. 60 six-cord, for example, and No. 60 three-cord are identical in size, though in reality No. 60 six-cord is formed of No. 120 yarns. It is relatively smoother, more even, and stronger than the three-cord grade. All sizes of six-cord threads are made of six strands, each of the latter being twice as fine as the number of the thread as designated by the label. Three-cord spool cotton is made of three strands of yarn, each of the same number as the thread.

Sizing. In textile manufacturing, sizing is the process of strengthening warp yarns by coating them with a preparation of starch, flour, etc., in order that they may withstand the weaving process without chafing or breaking. The operation of sizing is also often resorted to in finishing certain classes of cotton and linen fabrics, which are sized or dressed with various mixtures in order to create an appearance of weight and strength where these qualities do not exist, or, if present, only in a small degree. The object in sizing warp yarn before weaving is to enable that process to be performed with the minimum of threads breaking. Judicious sizing adds to the strength of the yarn by [Pg 146] filling up the spaces between the fibers, and by binding the loose ends on the outside of the thread to the main part. In order to accomplish this a number of ingredients are used in the size preparation, as no single material used alone gives satisfactory results. The filling up of the minute spaces in the yarns and the adhesion of the fibers produce a smooth thread with sufficient hardness to resist the continual chafing of the shuttles, reeds, and harnesses during the process of weaving. Flour and starch in a liquid state are used for this purpose, but owing to the liability to mildew, flour is not so much used as starch. Both of these materials, however, make the yarn brittle, and other ingredients are combined with them to overcome the brittleness. For a softener on heavy weight goods nothing has been found superior to good beef tallow. On light-weight goods the softener giving the most general satisfaction is paraffin.

When properly made the size preparation is a smooth mass of uniform consistence, free from lumps of any kind, and from all sediment and odor. Starch—the principal material which gives body to any size—requires the most careful treatment. It is first mixed with cold water into a smooth, creamy milk, which is slowly poured into the necessary quantity of boiling water until a clear, uniform paste is formed. Then the softeners are added, such as soaps, oils, and animal fats; next a small amount of gelatine or glue is stirred in and some form of preservative, usually chloride of zinc or salicylic acid. The mass is then thoroughly [Pg 147] stirred in tilted jacketed kettles with mechanical stirrers. The size may be applied to the yarn either hot or cold. When applied hot it penetrates into the interior, filling up every space between the fibers, binding all together, and forming a hard coating on the surface of the thread. A thorough washing or steaming serves to remove all the size from the woven fabric.

FINISHING ROOM

Cotton Finishing. Cotton fabrics, like other textiles, after leaving the loom must be subjected to various finishing processes so as to bring them into commercial condition. On piece-dyed goods part of the finishing is done before and part after the dyeing process. Each class of fabrics has definite finishing processes. In some cases weighting materials are added to the fabric [Pg 148] so as to hide more or less its actual construction. Cotton fabrics just from the loom present a soft and open structure, more so than other textiles. Therefore it is necessary to use proper finishing materials and processes which will fill up the openings or interstices as produced in the fabric by the interlacing of warp and filling, and at the same time give to the fabric a certain amount of stiffness. Of course this finish will disappear during wear or washing, it having been imparted to the fabric to bring the latter into a salable condition.

Cotton fabrics after weaving may be subjected to the following sub-processes of finishing:

Inspecting, Burling and Trimming, Bleaching, Washing, Scutching, Drying.

After the cloth leaves the loom it is brushed; then it passes over to the inspection table in an upward receding direction, so that the eye of the operator can readily detect imperfections. The ends of two or more pieces as coming from the loom are sewed into a string for convenient handling in the bleaching.

Bleaching. The object of bleaching is to free the cotton from its natural color. The ancient method of bleaching by exposure to the action of the sun's rays and frequent wetting has been superseded by a more complicated process involving the use of various chemicals. Pieces of cloth are tacked together (sewed) to form one continuous piece of from three to one thousand yards in length. The cloth is next passed over hot cylinders or a row of small gas jets to remove all the [Pg 149] fine, loose down from the surface. The goods are then washed and allowed to remain in a wet condition for a few hours, after which they are passed through milk of lime under heavy pressure, followed by rinsing in clear water. The goods are next "scoured" in water acidulated with hydrochloric acid, and boiled in a solution of soda, then washed as before in clear water. Next they are chlorined by being laid in a stone cistern containing a solution of chloride of lime and allowed to remain a few hours. This operation requires great care in the preparation of the chloride of lime, for if the smallest particle of undissolved bleaching powder is allowed to come in contact with and remain upon the cloth it is liable to produce holes. The goods are then boiled for four or five hours in a solution of carbonate of soda, after which they are washed. They are again chlorined as before and washed. The long strips are finally scoured in hydrochloric acid, washed, and well squeezed between metal rollers covered with cloth. After squeezing and drying, the cloth, if required for printing, needs no further operation, but if intended to be marketed in a white state, it must be finished, that is, starched or calendered.

Starching. The starch is applied to the cloth by means of rollers which dip into a vat containing the solution, while other rollers remove the excess. Sometimes the cloth is artificially weighted with fine clay or gypsum, the object being to render the cloth solid in appearance.

Calendering. The cloth is now put through the [Pg 150] calendering machine, the object of which is to give a perfectly smooth and even surface, and sometimes a superficial glaze; the common domestic smoothing iron may be regarded as a form of a calendering utensil. The cloth is first passed between the cylinders of a machine two, three, or four times, according to the finish desired. The calender finishes may be classed as dull, luster, glazed, watered or moire, and embossed. The calender always flattens and imparts a luster to the cloth passed through it. With considerable pressure between smooth rollers a soft, silky luster is given by equal flattening of all the threads. By passing two folds of the cloth at the same time between the rollers the threads of one make an impression upon the other, and give a wiry appearance. The iron rollers are sometimes made hollow for the purpose of admitting steam or gas in order to give a glaze finish. Embossing is produced by passing the cloth under heated metal rollers upon which are engraved suitable patterns, the effect of which is the reproduction of the pattern upon the surface of the cloth.

Mercerizing. This is a process of treating cotton yarn or fabrics with caustic soda and sulphuric acid whereby they are made stronger and heavier, and given a silky luster and feel. The luster produced upon cotton is due to two causes, the change in the structure of the fiber, and the removing of the outer skin of the fiber. The swelling of the fiber makes it rounder, so that the rays of light as they fall upon the surface are reflected instead of being absorbed. The quality and degree of [Pg 151] luster of mercerized cotton fabrics depends largely upon the grade of cotton used. The long-staple Egyptian and Sea Island cotton, so twisted as to leave the fibers as nearly loose and parallel as possible, show the best results. If the yarn is singed the result is a further improvement. Yarns and fabrics constructed of the ordinary grades of cotton cannot be mercerized to advantage. The cost of producing high-grade mercerized yarn is about three times that of an unmercerized yarn of the same count, spun from the commoner qualities of cotton.

Mercerized yarn is employed in almost every conceivable manner, not only in the manufacture of half-silk and half-wool fabrics, and in lustrous all-cotton tissues, but also in the production of figures and stripes of cotton goods having non-lustrous grounds. Mercerized yarn used in connection with silk is difficult to detect except by an expert eye.

Characteristics of a good piece of Cotton Cloth. A perfect cotton fiber has little convolutions in it which give the strong twist and spring to a good thread. In this respect the Sea Island cotton is the best. There are five things requisite for cotton cloth to be good, viz.:

1. The cloth must be made of good fiber, that is ripe and long.

2. The fiber must be carefully prepared. All the processes must be well performed—for the very fine thread fiber must be combed to remove poor fiber. The combing, however, is not always done.

[Pg 152] 3. The warp and woof threads must be in good proportion.

4. The cloth must be soft, so that it will not crease easily.

5. It must be carefully bleached—the chemicals used must not be strong.

[Pg 153]

CHAPTER XII

KNITTING

The art and process of forming fabrics by looping a single thread, either by hand with slender wires or by means of a machine provided with hooked needles, is called knitting. Crocheting is an analogous art, but differs from knitting in the fact that the separate loops are thrown off and finished by hand successively, whereas in knitting the whole series of loops which go to form one length or round are retained on one or more needles, while a new series is being formed on a separate needle. Netting is performed by knotting threads into meshes that cannot be unraveled, while knitting can be unraveled and the same thread applied to any other use. Knitting is really carried on without making knots; thus, the destruction of one loop threatens the destruction of the whole web, unless the meshes are reunited.

The principle of knitting is quite distinct from that of weaving. In the weaving of cloth the yarns of one system cross those of another system at right angles, thus producing a solid, firm texture. The great elasticity of any kind of texture produced by knitting is the chief feature that distinguishes hosiery from woven stuffs. The nature of the loop formed by the knitting needle [Pg 154] favors elongation and contraction without marring in the least the general structure of the goods. Builders of weavers' looms have at times endeavored to secure this elastic effect by certain manipulations of the mechanism of the loom, but as yet nothing approaching the product of the knitter has been made. The elastic feature of a knitted texture renders it peculiarly adapted for all classes and kinds of undergarments, for it not only fits the body snugly, but expands more readily than any other fabric of similar weight.

Knitting Machines. There are various machines for knitting. The circular knitting machine produces a circular web of various degrees of fineness, and in sizes ranging from a child's stocking to a man's No. 50 undershirt. The circular fabric made in this manner has to be cut up and joined together by some method to make a complete garment. The knitting frame for producing fashioned goods makes a flat strip, narrowing and widening it at certain places so as to conform to the shape of the foot, leg, or body. These strips then have to be joined by sewing or knitting to form a garment. Fashioning machines are indispensable for knitting the Niantic and French foot, and also for the production of stripes, fancy openwork, and lace hosiery.

[Pg 155]

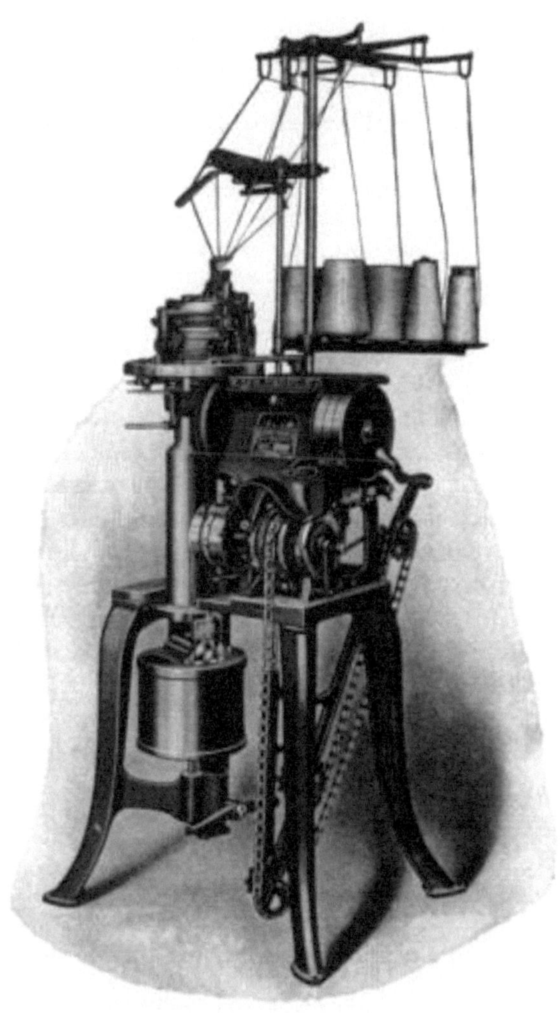

MACHINE FOR HOSIERY

[Pg 156] All plain machines of any class produce only plain knit-ted fabrics, while ribbed machines make only ribbed fabrics. Still, many garments in their make-up include both kinds of knitting; therefore, many machines produce only certain parts of particular garments. In the case of half-hose there is frequently a ribbed top, or in underwear a ribbed cuff, and these may be made either of circular web or full fashioned. In each case the ribbed portion is first knit and then transferred to a plain machine, and being placed upon the needles is worked on to the rest of the garment. In some instances the heel is made by the machine working the leg, though there are numerous knitters specially designed for turning out only this particular part.

Among other knitting machines in modern use are the drawers machine; machines for hose and half-hose with apparatus for making the instep, finishing off the toe, splicing or thickening the heels, etc.; machines for producing the bottoms or soles of hose separately, and also the instep separately; circular stocking machines for producing a tubular web afterwards cut into suitable lengths for all varieties of hose; circular sleeve machines, circular body machines, as well as circular web machines for making both body and sleeves of undershirts, jerseys, sweaters, etc. Special machines are also made for knitting both plain and ribbed plaited goods, that is, with both sides wool while the center is of cotton, or with a silk or worsted face on one side and the back of an inferior yarn. In the form of auxiliary appliances are produced many kinds of stitching machines; circular latch-needle machines for plain ribbed, mock seam, and striped goods; steam presses; hose rolling machines; hose cutting and welting machines, and many other accessories to hosiery manufacture.

KNITTING MACHINE FOR UNDERWEAR

At present fully one-third of the knit underwear used in this country is of the ribbed description. It [Pg 157] is made in all the materials that the older flat goods are composed of, including silk, silk mixtures, linen, wool, lisle, and cotton. Rib work is ordinarily stronger and [Pg 158] more lasting than plain. It is also invaluable for many purposes on account of its tendency to contract and expand in the direction of the circumference without altering its length. This feature makes it indispensable for tops to socks and wrist work for shirts, mittens, gloves, etc., and for the production of heavy garments such as cardigans and sweaters. The expense of knitting rib work is higher than plain knitting, owing to the fact that the machines cannot turn out so great a quantity within a given time.

The formation of the rib in knitted goods is unique in its principle. The effect is produced by reversing the stitch. In place of making the stitch work appear entirely upon one side of the fabric, as in plain work, the needles are so arranged that every alternate row, or two rows alternately, are reversed, thus making both sides alike. Plain work is done with a single bank of needles, while rib work requires two banks, the function of the second one being to pull and loop the yarn in an opposite direction, thus producing a thicker and more elastic web.

Double work in knitting consists merely in running two threads where one is commonly used. The work is done readily and with but little extra cost for labor. Coarser and heavier needles are required, also a wider gauge for the needle cylinder. Fancy effects in double work are produced by running two colors instead of one. The tendency is for one thread to twine about the other, thus making attractive double-and-twist work. Lumbermen's socks and like goods are often [Pg 159] knitted on this plan, though for the most part double work is for the heels, toes, and soles of ordinary hose.

Stripe Knitting. The process of striping knitted fabrics is accomplished automatically by a system of changing the yarns when delivered by the feeds. Circular machines knitting a tubular web cannot be utilized for this purpose, hence the work is done on fashioning or stocking frames. It has only been within recent years that makers of knitting machinery have been able to offer machines on which more than one kind of yarn could be knit at one time. There are now in use, however, machines that will readily knit several colors of yarn at the same time.

Knitting Cotton. A variety of loosely twisted, four-ply cotton yarn, dyed in various plain and mixed colors, employed for knitting hosiery, tidies, mats, etc., by hand. It is numbered from 8, coarse, to

20, fine, and commonly put up sixteen balls in a box, each box containing two pounds, manufacturer's weight.

Knitting Silk. A loosely twisted silk thread of domestic manufacture employed for knitting mittens, stockings, and other articles by hand. It is also much used for crochet work. Knitting silk is put up in the form of balls, each containing one-half ounce of thread. It is made in but two sizes, No. 300, coarse, and No. 500, fine; each ball of the former number contains 150 yards of silk; of the latter 250 yards. No. 500 is manufactured only in white, cream, and black; the No. 300 is fast dyed in a great variety of colors.

Hosiery Manufacture. According to the particular [Pg 160] method by which socks and stockings are made, of whatever kind, quality, or material, they are classed as cut goods, seamless, or full fashioned. Of the three methods of manufacturing the first named is the least expensive. Cut goods are made of round webbing knitted on what is called a circular knitting machine. The web has the appearance of a long roll of cloth about the width of a sock or stocking when pressed flat. The first operation consists in cutting off pieces the length of the stocking desired, these lengths, of course, being the same (unshaped) from end to end. The shaping of the leg is effected either by cutting out enough of the stocking from the calf to the heel to allow part to be sewn up and shaped to fit the ankle, or by shrinking. In the heeling room where the pieces next go, the cutters are furnished with gauges or patterns that indicate just where to make a slit for the insertion of the heel, generally of a different color. When the heel is sewn in, the stocking begins to assume its rightful shape. The toe is now put on and the stocking is practically finished. In the case of socks the final operation consists in attaching the ribbed top, which tends to draw the upper part of the leg together, thus causing it to assume a better shape. The final work includes scouring, dyeing, and shaping. The cost of making cut goods is less by a few cents per dozen than when knit seamless. While some very creditable hose are produced in this way, yet the existence of the heavy seam is an objection which confines them to the poorest class of trade. Cut goods are [Pg 161] made in all sizes and kinds for men, women, and children.

Seamless hose are made on a specially constructed machine which produces the entire stocking, but leaves the toe piece to be joined together by a looping attachment. On half-hose the leg is made the same size down to the ankle, but on ladies' hose the stocking is shaped somewhat in the machine. Seamless hose are not, strictly speaking, entirely seamless, inasmuch as all stockings made on a circular knitting machine must have a seam somewhere. There must be a beginning and an ending. In the case of the stocking the ending is at the toe, and the opening left can only be closed with a seam. In some mills this opening is automatically stitched together on special machines; in others, girls do it by hand with needle and thread. Neither by machine nor handwork can the opening be closed with exactly the same stitch as that made by the needles of the power knitter. However, the seam is of small proportions, and when the goods are scoured, pressed, and finished the presence of the seam is a minor item, as it neither incommodes the wearer nor mars the appearance of the stocking. Seamless goods are made in a great variety of qualities, ranging from cotton half-hose at fifty cents per dozen to the fine worsted stockings at $6.00 per dozen. A notable and very commendable feature of seamless hose is the socket-like shape of the heel, which fits that portion of the foot as though really fitted to it. As far as comfort and fit are concerned, the manufacture of seamless hosiery has now reached [Pg 162] such a degree of perfection as to bring it second only to the full-fashioned variety.

Full-fashioned hose are produced by means of complicated and expensive knitting frames, which automatically drop the requisite number of stitches at the ankle so as gradually to narrow the web down and give the stocking the natural shape of the leg. The toe is produced in the same way, and the shaping of heel and gusset is brought about in like manner. Hence, the goods are called full-fashioned, because so fashioned as to conform to the proportions of the leg and foot. Hose and underwear made by this method are knit in flat strips and then seamed either by hand or machine. Generally special machines are used, which take up and complete the selvedges, thus avoiding objectionable seams with raw edges.

The knitting frames used for making full-fashioned goods are large, intricate, expensive, and slow in operation; they are difficult to keep in order and require skilful operators. The largest ones knit from fourteen to eighteen stockings at once, using as many as four threads of different colors in the production of patterns. The first operation consists in knitting the leg down to the foot; then the legs are transferred by expert workmen to another frame which knits the foot. Next they go to another department where, with the aid of a special looping machine, the heels and toes are stitched together. Then the stockings or socks are handed over to expert women operators, who seam up the legs on a machine especially adapted for the purpose. After being sorted they are taken to be dyed, boarded, stitched, [Pg 163] dried, and finally subjected to heat and pressure to give them a finished appearance. It usually requires two weeks from the time the manufacturing operations begin, for a stocking to emerge from the factory in a finished form. Full-fashioned hose are made in all shades and grades of silk and cotton, in lisle thread, and in all kinds of cashmere, merino, and woolen goods. They are likewise knitted plain, ribbed, and with fancy stripes and embroidery effects. In the United States there are numerous important plants engaged in the production of full-fashioned goods, while large quantities are annually imported from Germany and France.

Finishing Process. When socks and stockings are taken off of the knitting machines they present an unfinished appearance, being loose, puckered, dirty, and generally shapeless. Scouring, dyeing, shaping, and pressing serve to improve their looks, and these finishing operations constitute a distinct branch of the industry. While still in a moist state the hose are shaped. This is effected by the use of forming-boards made of wood and about one-half of an inch in thickness. The sock or stocking is carefully stretched over the "form" while damp, and then placed in a heated chamber and allowed to dry. The goods assume the shape of the wooden "form," and will always hold it if the work has been carefully and thoroughly done. After they have been taken from the drying chamber and the boards removed the hose are pressed between heavy metal plates or rollers, looked over for defects, and when boxed or bundled are ready for market.

[Pg 164]

CHAPTER XIII

LACE

Lace. Lace is the name applied to an ornamental open work of threads of flax, cotton, silk, gold, or silver, and occasionally of mohair or aloe fiber. The latter are used by the peasants of Italy and Spain.

Lace consists of two parts, the ground and the flower. The threads may be looped, plaited, or twisted in one of three ways. First, with a needle, when the work is known as "needlepoint lace." Second, when bobbins, pins, and a pillow or cushion are used; this is called "pillow lace." Third, by machinery, when imitations of both point and pillow lace patterns are produced.

Special patterns for these laces date from the beginning of the sixteenth century. The early productions of the art had some analogy to weaving; the patterns were stiff and geometrical, sometimes cut out of linen or separately sewed and applied to the meshed surface, but more frequently they were darned in, the stitches being counted in, as in tapestry. This kind was known as darned netting. With the development of the renaissance of art, free flowing patterns and figure subjects were introduced and worked in.

Whether of needlepoint or pillow make, both the [Pg 165] ornament and the ground are produced by the lace maker. Needlepoint is made by first stitching the net with thread along the outline of a pattern drawn on paper or parchment, thus producing a skeleton thread pattern. This threadwork serves as a foundation for the different figures which are formed in the lace.

Bobbin or pillow lace more nearly resembles weaving. The threads are fixed upon a circular or square pillow, placed variously to suit the methods of manufacture in vogue in different countries. The object of using the pillow is to prevent too much handling of the lace. One end of each thread is fastened to the cushion with a pin, the main supply of thread being twined around a small bobbin of wood, bone, or ivory. The threads are twisted and plaited together by the lace maker, who throws the bobbins over and under each other. The operation is fairly simple, since children of eight or nine years of age can be trained to it successfully. It demands, however, considerable dexterity with the fingers.

The design for pillow lace must of course be adapted to the technical requirements of the process, and cannot therefore be the same as one for needlepoint, which has a better appearance and greater strength than pillow lace. For this reason it was in former times generally preferred for wear on occasions of state. On the other hand, pillow lace has the quality of charming suppleness, and for use in mantillas, veils, and fichus it is better than needlepoint, lending itself with delicate softness and graceful flexibility as a covering to the head and shoulders of women.

[Pg 166]

LACE TERMS DEFINED

Alençon (Point d').—Fine needlepoint lace with the ground of double-twist thread in a semi-net effect. Is usually worked with horsehair on the edges to give firmness to the cordonnet. Called after the city in France where it is made.

Allover.—Name for all wide laces used for flouncing, yokes, and entire waists. Usually the lace is over eighteen inches in width.

American Laces.—A general term formerly used to distinguish lace made in this country, the development of the industry having now rendered the term nearly obsolete.

Angleterre (Point d').—Fine Brussels pillow lace, distinguished by a rib of raised and plaited threads worked in the lace. Shown in floral, ornithological, and geometrical designs.

Antique.—Hand-made pillow lace of heavy linen thread in a large, open, rectangular knotted mesh. Used for curtains, bed sets, draperies.

Antwerp.—Bobbin lace, resembling early Alençon. Shows a "pot"—that is, a vase or basket effect—in the design.

Appliqué.—Any lace in which the body and the design are made separately. The body is usually silk and the design cotton or linen.

Appliqué Brussels.—Name sometimes given to Brussels appliqué laces.

Arabe (Point d').—Coarse bobbin lace made in Belgium and France as well as Arabia. Shows a large, bold pattern, cable edged, and is almost invariably in a deep écru tone. Used for curtains and draperies.

Arabian.—Same as above.

Argentine.—Similar to Alençon, the mesh being a trifle larger.

Arras.—Very strong, inexpensive, white bobbin lace, of simple pattern, somewhat resembling Mechlin. Distinguished by its light, single thread ground. Named after the city in France where it is made.

Aurillac.—Somewhat resembles Angleterre. Bobbin lace made in Aurillac, France.

Auvergne.—Any kind of bobbin lace made in Auvergne, France. Different makes and patterns.

Ave Maria.—A narrow edging lace.

Baby Lace.—Light and simple edging lace made in England.

[Pg 167] *Battenberg.*—Same as Renaissance. Designs confined to flower patterns.

Bayeux.—Bobbin lace, usually an imitation of Spanish point. Also a black, rich lace made in large pieces for shawls, head scarfs, etc.

Binche.—Fine pillow lace, without cordonnet. Ground resembles a spider-web with small dots. Made in Binche, Belgium.

Bisette.—Coarse, narrow French peasant lace in simple designs. Name often applied to cheap bordering laces.

Blonde.—So called, being originally a bobbin lace made of unbleached silk, though now shown in black, white, and colors. Made with two different sizes of thread; fine thread for the ground, coarse for the design. Usually takes some floral form. Very lustrous.

Bobbin Lace.—Imitation of pillow lace. Made in England and France.

Bobbinet.—The same.

Bone Lace. — An obsolete term once given to Honiton bobbin lace.

Bone Point Lace. — Applied to laces having no regular ground or mesh, such as Renaissance.

Border Lace. — Practically synonymous with edging.

Bourdon. — A machine lace made of both silk and cotton. Show scroll-like patterns cable-edged on a regular mesh. Usually dyed black, but sometimes bleached. The outline is of a heavy lustrous thread. Used chiefly for dress trimming and millinery.

Brettone. — Cheap narrow edging.

Bride Lace. — Lace with the pattern connected with brides. Same as bone point lace.

Brides. — Slender threads connecting different parts of a pattern.

Brussels Net. — Plain net made originally in Brussels, but now produced in all lace manufacturing countries.

Brussels Pillow. — Fine pillow lace with the patterns joined together by little loops on their edges.

Brussels Point. — Shows an open pattern, made partly in open, partly in closed, stitch, giving the appearance of shading.

Carrickmacross. — Tiny Irish cambric drawn work, appliqué on net.

Cartisane. — Guipure or passementerie made with thin silk or gilt-covered strips of parchment.

[Pg 168] *Chantilly.* — Pillow lace very similar to blonde. Comes from Chantilly, France. Made in both silk and cotton and usually seen in black. Non-lustrous, and looks as if made from black linen thread.

Chiffon Lace. — Chiffon embroidered in twist silk.

Cluny. — Coarse-thread bobbin lace, made in both linen and cotton. Shows a close-stitch pattern darned on an open ground. Used for dress trimmings and the manufacture of curtains.

Cork Lace. — A sweeping term used to designate all laces of Irish make.

Cotton Lace. — All lace made of cotton.

Craponne. — Cheap, stout thread furniture guipure.

Crochet Lace. — Any point lace made with the crochet hook.

Darned Lace. — A comprehensive term taking in all net effects with the pattern applied in needlework.

Devonshire Lace. — Lace made in this part of England, and especially Honiton imitation.

Dieppe. — Fine needlepoint lace made in Dieppe, France. Resembles Valenciennes. Made with a regular ground of squares of small meshes alternating with open squares upon which the pattern is applied in close stitch.

Duchesse. — Pillow lace with fine net ground with the patterns in raised work, volants, and the like.

Dutch Lace. — Practically a coarse Valenciennes.

English Point. — See **Angleterre** .

Escurial. — Heavy silk lace made in imitation of Rose point. Patterns outlined with cable edge.

Esprit (Point d'). — Dotted bobbinet with the dots either singly or in clusters.

Filet Lace. — Any lace made with a square mesh net.

Flemish Point. — Needlepoint lace made in Flanders.

Footing. — Simple insertion of Brussels net from one to three inches in width.

Galloon. — Irregular band with a fancy edge. Entire piece often in zigzag or scallop form.

Gaze (Point de). — Flemish point lace resembling point d'Alençon, though much softer, being without horsehair.

Gêne (Point de). — Openwork embroidery made on a wool ground which is afterwards eaten away by acid.

[Pg 169] *Genoa.* — Heavy lace made of aloe fiber. Another name for macramé.

Gimp. — See **Guipure** .

Gold Lace.—Gimp or braid covered with gold or imitation gold thread.

Grammont.—White pillow lace used for shawls and the like. Black silk lace nearly resembling blonde.

Guipure.—Fancy trimming of wire cord whipped round with silk or cotton threads, and the small patterns stitched together.

Guipure d'Art.—Linen net upon which raised intersecting patterns are worked.

Guipure de Flanders.—A pillow lace made separately, having flowers connected by bars and brides.

Hand Embroidered.—Heavy point lace, usually of Plauen manufacture, with fancy floral or other figures embroidered on the design.

Honiton.—English bobbin lace, famed for the beauty of its designs. Named for the city where it was first manufactured. Now made in Belgium, Holland, and France. Sprays sometimes made separately, and then worked on a net—Honiton appliqué.

Honiton Braid.—Narrow machine-made braid of ornamental oval figures connected by narrow bars. Used for collars, handkerchiefs, and tidies.

Honiton Guipure.—Large flower-pattern lace on very open ground, the sprays held together with brides or bars.

Imitation Lace.—A term used to designate any machine-made lace in contrast with hand-made.

Insertion.—Any narrow lace with a plain edge on either side that admits of its being inserted in a fabric.

Irish Crochet.—Heavy hand-made lace, remarkable for the beauty and distinctness of its patterns, and the startling whiteness of the linen thread used in its manufacture.

Irish Lace.—A general term used to designate all lace made by the Irish peasantry.

Irish Point.—Hybrid combination of appliqué, cut work, and embroidery on net with elaborate needle stitching in the higher grades.

Irish Trimming.—Simple, woven lace, used on white wear.

Knotted Lace. —Frequently referred to as knotting. A fancy weave of twisted and knotted threads in close imitation of some old hand laces.

[Pg 170] *Lille (Also Lile).* —French lace named after the town where it is made. Somewhat resembles Mechlin. Shows a very clear, light ground and is the most beautiful of all simple thread laces.

Limerick Lace. —A form of embroidery on net or muslin.

Luxeuil. —A general term for hand-made laces of Luxeuil, France. More specifically those of a stout, heavy nature. Used for tidies, curtains, draperies.

Macramé. —Knotted hand-made lace, made of a very heavy cord. Shown principally in geometrical designs. Very popular in deep écru.

Maline. —Fine silk net. Sometimes also applied to Mechlin lace with a diamond mesh.

Maltese. —Coarse machine-made cotton lace, resembling torchon. Has no regular ground, the patterns being usually connected with heavy stitch work.

Mechlin. —Light pillow lace with the pattern outlined by a fine but very distinct thread or cord. Real Mechlin generally has the ground pattern woven together, the latter running largely to flowers, buds, etc.

Medallion. —Single, detached pattern.

Medici. —Special kind of torchon edging, with one edge scalloped.

Mélange. —Hand-made silk pillow lace, showing a combination of conventional Chantilly with Spanish designs.

Mignonette. —Light bobbin lace, made in narrow strips. Resembles tulle.

Miracourt. —Sprig effects of bobbin-lace applied on net ground.

Mexican Drawnwork. —Little round medallions either single or in strips, the threads drawn to form a cartwheel. Mexican and Teneriffe drawnwork are practically the same. Machine imitations made in Nottingham, Calais, and St. Gall.

Motif.—See **Medallion** .

Nanduly.—South American fiber-lace, made by needle in small squares, which are afterward joined together. Design very beautiful and of remarkable durability.

Needlepoint Lace.—See **Point Lace** .

Normandy Lace.—See **Valenciennes** .

[Pg 171] *Nottingham.*—A general term including all the machine-made laces turned out in that great lace-producing center of England.

Oriental Lace.—Really an embroidery, being produced on the Schiffli machine, the pattern being then either cut or eaten out. Also applied to point d'Arabe and certain filet effects.

Oyah Lace.—A crocheted guipure shown in ornate patterns.

Passementerie.—A decorative edging or trimming, especially gimp or braid.

Picots.—Infinitesimal loops on brides and other strands.

Pillow Lace (Bobbin Lace).—Made on a pillow with bobbins and pins. Machine-made imitations retain the name.

Plauen.—Applied to all laces emanating from that section of Saxony and including imitations of nearly all point laces, which are embroidered on a wool ground, this being afterward dissolved in acid and the cotton or silk design left intact.

Point de Gaze.—Fine gauze-like needle-lace.

Point d'Irelande.—Coarse machine lace, made in imitation of real Venetian point.

Point de Milan.—A variety of guipure, having a ground of small meshes, and a pattern consisting of bold, flowing scroll devices.

Point de Paris.—A variety of cheap machine lace, cotton, of simple design.

Point Kant.—Flemish pillow lace, with a net ground and the design running largely to "pot" effects—pot lace.

Point Lace.—Lace made by hand with needle and single thread. Needlepoint the same. Point d'Alençon, point de Venise, etc., are all variations of point lace and will be found classified under their initials.

Point Plat.—Point lace without raised design.

Renaissance.—Modern lace, made of narrow tape or braid formed into patterns, held together by brides, the brides forming subsidiary designs. Battenberg is the same thing.

Repoussé.—Applied to the design, being a pattern that has the effect of being stamped in.

Rococo.—Italian lace, bearing the rococo design.

Rose Point.—See **Venetian point**.

Seaming Lace.—Narrow, openwork insertion.

[Pg 172] *Seville.*—Variety of torchon.

Spanish Lace.—A comprehensive term. Convent-made, needle-point lace. Cut drawnwork effects, also convent-made. Needlepoint lace in large squares. Black silk lace in floral designs.

Spanish Point.—Ancient embroidery of gold, silver, and silk passementerie.

Swiss Lace.—Swiss embroidered net in imitation of Brussels.

Tambour.—Variety of Limerick.

Tape Lace.—Hand-made needle lace, similar to Renaissance.

Thread Lace.—Made of linen thread, as distinguished from cotton and silk laces.

Torchon.—Coarse, open bobbin lace of stout but loosely twisted thread in very simple patterns. Much seen in imitations, usually in narrow widths.

Van Dyke Points.—Applied to laces with a border made in large points.

Valenciennes.—Commonly called Val. Bobbin lace, seen mostly in cheap insertions and in the form of narrow edgings.

Venetian Point.—Point de Venise. Needlepoint lace in floral pattern with the designs very close together and connected by brides ornamented with picots.

Wood Fiber.—Applied to all laces made of wood silk.

Yak.—Machine-made worsted lace. Used for trimming for shawls, petticoats, and undergarments.

Youghal.—Needlepoint lace of coarse thread, made exclusively in Ireland.

Ypres.—Bobbin lace, somewhat coarser than Val.

[Pg 173]

CHAPTER XIV

COTTON FABRICS [16]

Albatross. Cotton albatross cloth is a fabric made in imitation of a worsted fabric of the same name. It has a fleecy surface. The name is taken from the bird whose downy breast the finish of the fabric resembles. The warp is usually 28s cotton, the filling 36s cotton. It is a plain weave. Filling and warp count 48 picks per inch. The goods are finished by being burled, sheared, washed, singed, dyed, rinsed, dried, and pressed, care being taken not to press too hard. Sometimes singeing is omitted. Albatross cloth is generally in white, black, or solid colors. It is not often printed. It is light in weight, and is used for dress goods.

Awning. A cotton cloth used as a cover to shelter from sun rays.

Batiste. Batiste is of French origin, and is a light, transparent cloth, made from a fine quality of combed cotton yarn. There is a gradual variation in quality ranging from a comparatively coarse to a very fine fabric. The variety of qualities will suggest some idea of the utility of the fabric. Its uses are even more varied than are the qualities. The finer grades are [Pg 174] used for dress goods and all kinds of lingerie for summer wear, etc., while the cheaper grades are used for linings in washable and unwashable shirt waists. Batiste is woven in the gray, that is, with yarn direct from the spinning frame, with the exception that the warp yarn is well sized, in order to stand better the strain to which it is subjected during the weaving process.

Bourrette. A light weight, single cloth fabric, with two-ply cotton warp and wool or a combination of cotton and shoddy filling, made with the plain weave and in appearance a semi-rough-faced woolen fabric with fancy effects in twist scattered about it. It is used principally for ladies' fall suitings.

Bedford Cord. This is one of the most popular types of fabrics, the distinguishing effect being a line or cord running lengthwise of the cloth, the cord being more or less prominent. The cloth is made of cotton, or sometimes of worsted. The face effect of the Bedford cord is generally plain. Occasionally twill-faced cords are used. The cords vary in width from about one twentieth to one quarter of an inch. To get extra weight without altering the appearance of the face, extra warp yarns, termed wadding ends, are inserted between the face weave and the filling, floating at the back of the rib. When these wadding ends are coarse, they give a pronounced rounded appearance to the cord. They run from 88 to 156 picks to an inch.

Buckram. Buckram is derived from Bokhara. It may be described as a coarse, glue-sized fabric, and is made of cotton, hemp, linen, or cotton and hair (coarse) [Pg 175] yarns, usually from 10s to 25s. Made of a double cloth warp, 22s cotton, 34 picks to the inch, for the face or top fabric 1/12's [17]; weight from loom 2.22 ozs. per yard. Bottom fabric 1/12's cotton; filling 1/16's cotton; 12 picks to the inch. Weight per yard, 1.8 ounces. These fabrics depend a great deal on the finishing. The men's wear requires less sizing on account of the hair it contains. The goods are piece dyed. Buckram is used principally for stiffening garments, and to give them shape or form. It is placed between the lining and the surface cloth of the garment in particular parts, such as the lapels, etc. It is used in the millinery trade, and is made into hats. Millinery buckram is sized two or three times.

Calico takes its name from Calicut, a city in India, where cloth was first printed. The majority of inexpensive cotton fabrics are constructed on the one up, one down system, or plain weave. Calico is no exception to this rule. The printed designs on calicoes may be somewhat elaborate or they may be simple geometrical figures. In order, however, to comply with the true principles of art, such fabrics as calicoes should have but simple geometrical figures for their ornamental features. New styles and combinations of colors are produced every month and faster and lighter color printed each season. Most of the designs for calicoes and cotton cloth printing are made in Paris. At present the steam styles are most prominent; they are the fastest and lightest to be obtained. Calico is a printed cloth, [Pg 176] the printing being done by a printing machine which has a rotating impression cylinder on which the design has been stamped or cut out. The cloth in passing through the machine comes in contact with the impression cylinder. The cylinder revolving in a color

154

trough takes up the color and leaves the impression of the design on the cloth. Calicoes may be seen in almost any color. The printing machine is capable of printing several colors in one design. Calicoes, however, are usually in two colors, that is, one color for ground and the other for figure. The ground color in most cases is effected by dyeing the cloth in some solid color. After the cloth is dyed the design is printed on it. The cloth, after it comes from the loom, is singed and bleached, then sheared and brushed to take away all the lint, and then sent to the dye house. The first process there is to boil it, after which it is immersed in the dye tub. Calicoes are usually given what may be termed a "cheap cotton dye." By "cheap cotton dye" is meant that the colors are not fast, but will run or fade when subjected to water. After the fabric is dyed, it is given to the printer, who ornaments the face of the cloth with some geometrical design; then it is practically ready for the merchant. After printing, the cloth is dried and steamed to fix the color, afterwards soaped, washed, finished, and folded. The printing machine turns out about 400 to 800 fifty-yard pieces a day. Calico is used for inexpensive dresses, shirtwaists, wrappers, etc.

Cambric. Cambric is a heavy, glazed cotton fabric with a smooth finish. It was first made in Cambrai, [Pg 177] France. It has a plain weave and a width of thirty-six inches. Cambrics are dyed in a jig machine. After dyeing they are run through a mangle containing the sizing substance, then dried, dampened, and run through a calender machine. The glossy effect is obtained in this last finishing process. Cambric is used for shirtwaists, dress goods, etc. The finer grades are made from hard twisted cotton of good quality.

Canvas. This is a term applied to heavy, plain weave cloths made with ply cotton yarn. They are used for mail bags, covering for boats, etc.

Chambray. Chambray is a staple fabric of many years standing, being next in rank among cotton goods after the better grade of gingham. Chambray is a light-weight single cloth fabric that is always woven with a plain weave, and always has a white selvedge. In effect it is a cloth having but one color in the warp, and woven with a white filling, this combination producing a solid color effect, the white filling reducing any harshness of warp color in the cloth.

It is composed of one warp and one filling, either all cotton, cotton and silk, or all silk. It is twenty-seven to thirty inches in width and single 30s cotton warp to single 60s silk, the count of yarn being governed by the weight per yard desired. The weight per finished yard is two to three and one-half ounces. Good colors for the warp are navy blue, dark brown, lavender, black, nile green, etc. When made of cotton warp and filling the fabric receives a regular gingham finish. The loom width can be restored by tentering or running the [Pg 178] goods over a machine fitted underneath with a series of coils of steam pipe. The top of this machine is fitted with an endless chain with a row of steel needles standing erect upon its face. Chains are adjusted to the width desired, and as the machine runs, both selvedges are caught by the needles and the cloth stretched to the required width.

Cheese Cloth. This is a thin cotton fabric of light weight and low counts of yarn, which ranks among the cheapest in cotton goods. It is used for innumerable purposes. The bleached fabric is used for wrapping cheese and butter after they are pressed. It is also much in demand for bunting for festival occasions, light curtains, masquerade dresses, etc. When used for bunting, draperies, and the like it is usually in colors, red, blue, cream, and yellow seeming to have the greatest demand. The weave is one and one or plain weave.

Chiné. Sometimes applied to glacé silk, or cotton two-toned effects. The name is French, meaning woven so as to have a mottled effect.

Chintz. Printed cotton cloth, with large, many-colored designs, used for furniture covering. The Hindoo wears it as a body covering. Chintz is the Hindoo word meaning variegated.

Cotton Flannel. Napped cotton flannel. Made first for trade in Canton, China.

Crash. A plain fabric for outing suits, towels, etc.

Crêpe. A fine, thin fabric of open texture made of cotton.

Crepon. Large designs in figured crêpe. The name [Pg 179] applies to the crispiness of the finish and is from the French word *crêper*, to make crisp.

Cretonne. Heavy cotton cloth printed in large designs, for drapery and furniture use. Cretonne was a Frenchman who first made the cloth.

Crinoline. Crinoline is a fabric composed of cotton warp, horsehair filling, or all cotton yarns. It is sold in varying widths, and is used by tailors and dress-makers in stiffening clothing. It is a cheap cloth of low texture and simple construction, the distinguishing feature being the stiff finish with either a dull or highly glazed face on the cloth.

Damask. A cloth of silk and cotton, silk and linen, silk and wool, or all linen in flowered or geometrical designs for drapery or table covering. The weaves used are mostly twills and sateens. It takes its name from Damascus, where it was first made.

Denim. This is a strong fabric usually made with a two up and one down twill. It is used for overalls, furniture covering, and floor covering.

Diaper. A figured cotton or linen fabric, which gets its name from the Greek *diapron*, meaning figured. It is generally of good quality as it is subject to excessive washing.

Dimity. A light-weight cotton fabric, the distinguishing feature of which is the cords or ribs running warpwise through the cloth, and produced by doubling the warp threads in either heddle or reed in sufficient quantity to form the rib desired. The name is from a Greek word meaning two-threaded. Dimity is a ladies' [Pg 180] summer dress fabric, and is made of regular cotton yarn, from 1/60's to the finest counts in both warp and filling. It is made in both white and colors, solid white being used in the most expensive grades. Colors are often printed upon the face of the fabric after it has been woven in the white.

Domet. This cloth is napped similar to a cotton flannel. It is used for shirts, pajamas, etc., and made with bright colored stripes and check patterns. The name is from domestic, home made.

Duck. Duck is a heavy single cloth fabric made of coarse two-ply yarn and of a plain weave. It derives its name from its resemblance to a duck's skin. It is of a lighter weight than canvas. In finishing duck is taken from the loom and washed and sized, then dried and pressed. If a fancy solid color is desired the goods are dyed in the piece after the first washing. Duck is used in the manufacture of sails, tents, car curtains, and for any purpose requiring a good water-tight fabric, which will withstand rough usage. Duck has a stiff hard feel, and excellent wearing qualities. The lighter weights are used for ladies' shirtwaist suits, men's white trousers, etc.

Drill. A cotton fabric of medium weight generally made with the two up and one down twill. It is extensively used for shoe linings.

Eolienne is the name applied to a fine dress fabric characterized by having the filling of a much coarser count than the warp, thus producing a corded effect across the breadth of the goods. This class of goods [Pg 181] is made up of a raw silk warp and either cotton or worsted filling, with the warp ends per inch greatly in excess of picks per inch. The goods are made up in gray, then dyed in the piece in any color the trade desires. The darker shades find most favor for fall and winter use, while the lighter shades are preferred for summer wear. The width is from twenty-seven to fifty inches, and the price per yard varies from 85 cents to $1.25.

Etamine. An etamine is a thin, glossy fabric used principally for women's dress goods. Being a common and popular material for summer wear, it is usually made as a piece-dyed fabric. A good reason for making it piece-dyed is that this method is much cheaper than if the yarn is dyed previous to the weaving. Etamines were originally made with worsted yarns, which of course are more expensive; however, if a good quality of cotton is used, there is little difference in appearance between worsted and cotton etamine. The difference is chiefly in the wearing quality, worsted being more durable. The principal characteristic of an etamine is a crisp, glossy, and open structure.

Flannelette is a narrow, light-weight fabric composed of all cotton yarn, the filling being soft spun to permit of the raising of a very slight nap on the back of the goods. The cloth is woven with bleached yarn (warp and filling), the color effects being afterwards printed upon the face of the goods by the printing machine. Flannelette is made with simple one or two colored stripe patterns, either black and white or indigo blue and [Pg 182] white, and in imitation of a Jacquard pattern. The finished fabrics are sold by the retailer at from eight cents to twelve and one-half cents per yard, are twenty-seven inches wide, and are used very extensively in the manufacture of ladies' wrappers, kimonos, etc., for house wear.

Fustian. A corded fabric made on the order of corduroy and used in England for trouserings, etc. First made at Fustat, a town on the Nile, near Cairo. Velveteen and cordings in the lower, coarser grades were sometimes called Fustian.

Galatea Cloth. Galatea cloth has been somewhat in demand in recent years by women requiring serviceable and neat-appearing cotton fabrics at a medium price. It is usually finished twenty-seven inches wide and retails at fourteen cents to twenty cents per yard. It is shown in plain colors as well as in figures, and in dotted and striped designs on white and colored grounds. The patterns are obtained by printing. Some manufacturers have found that they can take a standard type of fabric and extend its use by varying the process of finishing. The base of the cloth—that is, the fabric previous to dyeing or printing or bleaching—is nothing more than an ordinary 5-end warp sateen of fair quality.

Gauze. A veiling net, made in Gaza in Palestine.

Gingham. Gingham is a single cloth composed entirely of cotton, and always woven with a plain weave. It is yarn-dyed in stripes or checks and was originally of Indian make. It is the most widely known fabric on the market and is made in various grades, having from fifty [Pg 183] to seventy-six ends per inch in the reed, and of 1/26's to 1/40's cotton yarns in both warp and filling. It is a wash fabric, made in both check and plaid patterns into which an almost unlimited variety of color combinations are introduced. Ginghams are made with from two colors, warp and filling, to eight colors in warp and six in filling. Ginghams are used most commonly in the manufacture of ladies' and children's summer dresses and aprons.

Italian Cloth is a light, glossy fabric made from cotton and worsted, cotton and wool, cotton and mohair, and all cotton. It is used for linings for the heavier styles of ladies' dresses, also for underskirts, fancy pillow backs, etc. The cloth is woven in the gray undyed yarns. In the finer grades the warp is sized so as to facilitate the weaving process.

Jaconet. A thin cotton fabric, heavier than cambric. If properly made one side is glazed. Derived from the French word *jaconas*.

Khaki. Twilled cotton cloth of a brown dust color, first used for men's clothing in India. The word *khaki* is Indian for earth, or dust-colored.

Lawn. Lawn is a light-weight single cloth wash fabric, weighing from one and one fourth to two and one fourth ounces per yard, and in widths from thirty-six to forty inches finished. It is composed of all cotton yarns (bleached) from 1/40's to 1/100's, and is always woven with a plain weave, one up, one down. The name is from Laon, a place near Rheims, France, where lawn was extensively made. Plain lawn is made of solid white or bleached yarn in both warp and filling. The fancier [Pg 184] grades, or those having color effects, are produced by printing vines, floral stripes, small flowers, etc., in bright colors in scattered effects on the face of the goods. The patterns are always printed, never woven. Lawn, when finished, should have a soft, smooth feel. Therefore the finishing process includes brushing, very light starching or sizing, then calendering or pressing. Lawns have to be handled carefully in the bleaching process, starched with an ordinary starch mangle (the sizing containing a little blueing), finished on the Stenter machine, and dried with hot air. Lawns are often tinted light shades of blue, pink, cream, pearl, green, and other light tints, with the direct colors added to the starch. It is used principally in the manufacture of ladies' and children's summer dresses, sash curtains, etc.

Lingerie. This relates to all sorts of ladies' and children's undergarments, such as skirts, underskirts, infants' short dresses, chemises, night robes, drawers, corset covers, etc.

Linon is a fine, closely woven plain fabric, well known for its excellent wearing and washing qualities. It is made from combed cotton yarns of long-stapled stocks to resemble as closely as possible fine linen fabrics. The cloth structure is firmly made in the loom.

Long Cloth is a fine cotton fabric of superior quality, made with a fine grade of cotton yarn of medium twist. Originally the fabric was manufactured in England, and subsequently imitated in the United States. The fabric is used for infants' long dresses, from which it derives its name, and for lingerie. Long cloth to some [Pg 185] extent resembles batiste, fine muslins, India linen, and cambric. It is distinguished from these fabrics by the closeness of its weave, and when finished the fabric possesses a whiter appearance, due to the closeness of the weave and the soft twist of the yarn. It is not used as a dress fabric, chiefly because of its finished appearance, which is similar in all respects to fabrics which we have been accustomed to see used solely for lingerie, nightgowns, etc.

Madras is a light-weight single cloth fabric, composed of all cotton or cotton and silk, and has excellent wearing qualities. It was at first a light-colored checked or striped plain-faced cotton-silk fabric, made in Madras, India, for sailors' head-dress. It is twenty-seven inches wide, and is made of varying grades, weighing from two to three ounces per yard, and is used at all seasons of the year. It is used by ladies for summer skirts, shirtwaists, suits, etc., and by men in shirts. It is known by the white and colored narrow-stripe warp effects, and is made of cotton yarns ranging from 1/26 to 1/80 warp and filling, and from 50 to 100 or more ends per inch. The utility of madras for nearly all classes of people permits the greatest scope in creating both harmonious and contrasting color and weave combinations.

The colors most in demand in this fabric are rich and delicate shades of blue, rose, green, linen, tan, lavender, and bright red; for prominent hair-line effects black, navy blue, dark green, royal blue, and cherry red. Good fast color is necessary as it is a wash fabric. If inferior colors are used, they will surely spread during [Pg 186] the finishing processes, and will cause a clouded stripe where a distinct one was intended.

Moreen. Heavy mohair, cotton, or silk and cotton cloth, with worsted or moire face. The making of moreen is interesting. The undyed cloth is placed in a trough in as many layers as will take the finish. This finish is imparted to the cloth by placing between the layers sheets of manila paper; the contents of the trough are then saturated with water; a heavy weighted roller is then passed over the wetted paper and cloth, the movement of the roller giving the cloth a watered face. It can then be dyed and refinished. The design or marking of moreen is different on every piece. Moreen was at first made for upholstery and drapery use. It was found to give a rustling sound similar to silk, so was taken up for underskirts. The name is from the French *moire*, meaning watering.

Mull. A soft cotton muslin of fine quality, made first in India, later in Switzerland. The name in Hindoo is *mal*, meaning soft, pliable.

Mummy. A plain weave of flax or linen yarn. Originally the winding cloth of the Egyptian mummified dead.

Muslin. A fine cotton cloth of plain weave originally made in Mosul, a city on the banks of the Tigris, in Asia.

Nainsook. Nainsook is a light cotton fabric utilized for various purposes, such as infants' clothes, women's dress goods, lingerie, half curtains, etc. The striped and plaid nainsook are used for the same purposes. When the fabric is required for lingerie and infants' [Pg 187] clothes the English fabric is selected because of its softness. When intended for dress or curtain fabric, the French-finished fabric is chosen. The latter finish consists of slightly stiffening and calendering the cloth. The fabric may be distinguished from fine lawns, fine batiste, and fine cambric by the fact that it has not as firm construction or as much body, and the finish is not as smooth or as stiff, but inclines to softness, as the fabric has not the body to retain the finishing material.

Organdie. An organdie may be defined as a fine, translucent muslin used exclusively for dress goods. The fabric is made in a variety of qualities as regards the counts of yarn used, and in a variety of widths ranging from eighteen to sixty inches. The plain organdie is popular in pure white, although considerable quantities are dyed in the solid colors, pale blue, pink, etc., while the figured organdies are usually bleached pure white, then printed with small floral designs. The printed design is in from two to four colors, and in delicate shades in conformity with the material. Organdie considered in relation to cost as wearing material is rather expensive. The reason for this is that it has a finish peculiar to itself, so that when washed it does not have the same appearance as before. It loses its crisp feeling altogether.

Osnaburg. A coarse cloth of flax and tow, made in America of cotton, in checks or plaids, and used for furniture covering and mattress making. The town of Osnaburg, in Germany, made the fabric first.

[Pg 188] **Percale.** Percale is a closely woven fabric made with a good quality of cotton yarn. The finer qualities are used for handkerchiefs, aprons, etc., and when used for these purposes are not printed, but bleached after the fabric comes from the loom. Percale is chiefly used for dress fabrics, and when used for this purpose is generally printed on one side with geometrical figures, generally black, although other colors may be seen. The fabric is bleached before it is subjected to the printing operations.

Percaline. Percaline is a highly finished and dressed percale. The first process to which the cloth is subjected is to boil it off, that is, to soak it in boiling water so as to relieve it from foreign matter that it may have gathered during the weaving, and at the same time to prepare it for dyeing. After dyeing it is sized to stiffen it, and also to increase the gloss on the cloth. After sizing it is ready for the calender. In order to give it the highest gloss the cloth is doubled lengthwise or the pieces are put together back to back, and as it passes through the rolls it is wet by steam, the rolls being well heated and tightly set together. Percaline is used chiefly for feminine wearing apparel, principally for linings, petticoats, etc. These purposes require that the cloth shall be solid color, the darker colors being preferred, as blue, green, and black. Sometimes it is seen in lighter shades of brown and tan. The most attention is given to the finishing process.

Piqué. Piqué is a heavy cotton material woven in corded or figured effects. The goods are used for such [Pg 189] purposes as ladies' tailor-made suits, vestings, shirt fronts, cravats, bedspreads, and the like. It was originally woven in diamond-shaped designs to imitate quilting. The name is French for quilting. The plainest and most common fabrics of piqué are those in which the pattern consists of straight cords extending across the cloth in the direction of the weft. In the construction of these fabrics, both a face and back warp are required, and the cords are produced by all the back warp threads being raised at intervals of six, eight, or more picks over two or more picks of the face cloth, which has a tendency to draw down on the surface of the fabric. The goods are always woven white and no colors are ever used. The face warp threads are generally finer than the back warp threads, and are in the proportion of two threads for the face and one thread for the back. On the heavier and better grades of piqué coarse picks called wadding are used to increase the weight, and also to give more prominence to the cord effect. They are introduced between the face and back cloths. In the lightest and cheapest grades neither any wadding nor back picks are used. In this case the back warp threads float on the back of the fabric except when raising over the face picks to form the cord. In the figured piqué the binding of the back warp threads into the face cloth is not done in straight lines as in plain piqué, but the binding points are introduced so as to form figures. These fabrics are woven in the white, and the figures are purely the result of binding the face and back cloths together.

[Pg 190] **Poplin.** Poplin or popeline is a name given to a class of goods distinguished by a rib or cord effect running width way of the piece. It referred originally to a fabric having a silk warp and a figure of wool filling heavier than the warp. At the present time it refers more to a ribbed fabric than to one made from any particular combination of materials. Cotton poplin is usually made with a plain weave, the rep effect being obtained either by using a fine warp as compared with the filling, or a large number of ends as compared with picks per inch on both. Irish poplin is a light-weight variety of poplin, sometimes called single poplin, and is celebrated for its uniformly fine and excellent wearing qualities. It is principally made in Dublin.

Plumetis. Sheer cotton or woolen cloth having raised dots or figures in relief on plain ground. The design shows a feathery effect, as in embroidery tambour. The name is French for this kind of embroidery, and is derived from *plume*, French for feather.

Rep. A fabric having a surface of a cord-like appearance. The name is probably corrupted from rib. It is used in making shirt-waists and skirts.

Sateen. Twilled cotton cloth of light weight, finished to imitate silk satin. There are two kinds, viz., warp sateen and filling sateen.

Scrim. Open mesh weave of cotton or linen for curtains and linings. The name is from scrimp, referring to economy in weaving.

Silesia is a light-weight single cloth fabric, having a rather high texture, and weighing about three ounces [Pg 191] per yard. It is composed of all cotton yarn, and is used principally as a lining for ladies' and men's clothing. Silesia is woven of yarn in the gray state, and is dyed in the piece in such colors as black, dark blue, brown, drab, slate, steel, etc. An important feature is the highly glazed or polished face of the goods, which is due to the action of the heated roller in the calendering machine upon the sizing.

Souffle. The largest designs of crepon show a raised or puffed appearance. Souffle is from the French and means puffed.

Swiss. From Switzerland, where the plain Swiss net and figured cambric is a specialty in the St. Gall district.

Tape. Tape is a narrow fabric composed either of cotton or linen yarns in warp and filling, and usually made with a point or broken twill weave, the break in the weave occurring in the center of the tape, and the twill lines running in a right- and left-hand direction. It is used as a trimming in the manufacture of clothing, also as a binding in innumerable cases, and is sold by the roll, each roll containing a certain number of yards. It is made of all bleached and of regular yarns about 1/26's to 1/30's and 1/40's cotton.

Tarletan. An open mesh of coarse cotton, used mostly in fruit packing, sometimes for dress and drapery. The name is from *tarlantanna*, Milanese for coarse weave of linen and wool.

Terry Cloth or Turkish Toweling is a cotton pile fabric. It is woven in such a way as to permit the [Pg 192] forming of a series of loops on each side of the cloth in regular order. After leaving the loom each piece is laid separately in the bleaching kier. Then the goods are dried on a tenter frame, given a light starching to add weight, run through a rubber rolled mangle and again dried on a tenter frame. This cloth is used in the manufacture of towels, Turkish bath robes, etc. Turkish toweling is the original terry. The name is from the French *tirer*, to draw or pull.

Zephyr Gingham is the finest grade of gingham made and is a light-weight cotton fabric, composed of 1/40's to 1/60's cotton warp and filling yarns. It is woven with either the plain weave or a small all-over dobby effect. It is made in attractive patterns by using good fast colors in warp and filling, and as a cloth has excellent wearing qualities.

FOOTNOTES:

[16] This information is from the leading authority, "The Cotton Fabrics Glossary," published by the *American Wool and Cotton Reporter*, Boston, Mass., and is reprinted here through the kindness of Mr. Frank P. Bennett.

[17] 1/12's cotton signifies single cotton yarn of 12's. 2/12's cotton signifies two sets of single cotton yarn of 12's twisted together.

[Pg 193]

CHAPTER XV

FLAX

Flax. Flax or linen occupies the first position in the group of stem fibers, [18] being not only the oldest, but next to cotton the most important vegetable spinning material known. Its value is increased by the fact that the flax plant readily adapts itself to various conditions of soil and climate, and in consequence has gained access to northerly districts and cool highlands. Although flax has lost some of its importance from the successful competition of cotton, nevertheless it still forms one of the chief articles of an industry which merits all the care bestowed on its cultivation and proves highly profitable.

The Physical Structure of Flax. Flax, when seen under the microscope, looks like a long, cylindrical tube of uniform thickness, with lumina so small as to be visible only as straight black lines lengthwise of the fiber, and frequently exhibits small transverse cracks. It is never twisted like cotton fiber. Its color varies from pale yellow to steel gray or greenish tints. The [Pg 194] difference in color is due chiefly to the process of "retting." Its average length is about twenty inches, and its tensile strength is superior to that of cotton. It will absorb moisture, 12 per cent being the standard allowance made.

Flax is used for making linen thread and cloth, yarn, twist, string fabric, and lace. In its composition it is almost purely an unlignified cellulose, and its specific gravity is 1.5.

Flax is a better conductor of heat than cotton, hence linen goods always feel colder than cotton goods.

Russia produces more than one-half the world's supply of flax, but that from Belgium and Ireland is of the best quality. Italy, France, Holland, and Egypt are other important producers. The plant is an annual, of delicate structure, and is gathered just before it is ripe, the proper time being indicated by the changing of the color from green to brown. At the time of gathering the whole plant is uprooted, dried on the ground, and finally rippled with iron combs, to separate the stalks from the leaves, lateral shoots, and seeds.

The best fiber amounts to about 75 per cent of the stalk. To separate this valuable commercial product from the woody matter the stalks are first subjected to a process termed retting, which is steeping them in water until they are quite soft. Then follow the mechanical processes to further the production of the fiber and free it from all useless matter.

These are as follows:

1. Crushing or Beating. This consists of breaking [Pg 195] the woody matter with the aid of mallets or in stamping mills.

2. Breaking. This is passing the stalks through a series of horizontal rollers to break further the woody matter and at the same time separate the greater part of it from the fiber.

3. Scutching. The object of this process is to remove completely the woody matter, and it is done by means of rapidly revolving wooden arms or blades, which beat the firmly held flax until it is sufficiently cleaned and separated.

4. Hackling. The scutched flax is drawn through iron combs which still further open the fiber. Fineness of fiber depends upon the number of times it is hackled, each time with a finer and finer instrument, which secures the different degrees of subdivision. Then the fibers are sorted and classified as to length and quality and laid in parallel forms ready for spinning and manufacture into linen.

PULLING FLAX IN MINNESOTA

[Pg 196] **Bleaching.** Linen is bleached in the form of yarn, thread, and cloth. This is a difficult and long process owing to the large amount of natural impurities present in flax fiber, and the difficulty of removing or dissolving them. Bleaching is now done as a rule by chemical processes, and when chemicals are used great care must be taken about their strength and about the time the cloth is allowed to remain in them. In olden times sour buttermilk was applied to linen and rubbed in, and then bleaching was finished out of doors by sun and rain. "Unbleached" linen is treated in the same way as bleached, only the process is not carried to such an extent. In Ireland, famous for its bleaching, chemicals are used in the earlier stages of this process, and then fine linens are spread out on the grass to improve their color, and to purge them completely of any chemicals used. After bleaching, linen is washed, dried, starched, and put through heavy machines to give it a glossy finish, and it is then made up in pieces for sale.

Characteristics of Good Linen. Linen is noted for its smoothness of texture, its brilliancy — which laundering increases — its wearing qualities, and its exquisite freshness. The celebrated Irish linen is the most valuable staple in the market, and on account of its fineness and strength, and particularly its bright color, it attains an unapproachable excellence because the best processes are used throughout the entire manufacture. Linen is less elastic and pliable than cotton and bleaches and dyes readily.

Flax from all countries is woven into table linen, [Pg 197] though very fine linen must have carefully prepared fiber. Linen should be soft, yielding, and elastic, with almost a leathery feel. Fineness of linen does not always determine good wearing qualities.

Good linen ranges in price from 75 cents to $3.00. Irish linen has a good bleach. French and Belgian linens, while fine in thread, are not as serviceable as Irish linen. Germany makes a good wearing linen, but not a large variety of patterns. Scotch linens are now used more than other kinds.

STACKS OF FLAX IN BELGIUM
Copyright by Underwood & Underwood, N. Y.

Sources of Flax

Russia,
Holland,
Belgium,
Germany,
Ireland,
Canada,
U. S. (for seed only).

[Pg 198]

Sources of Manufactured Linens

170

Scotland,
Ireland,
Germany,
Austria,
Belgium,
France,
Russia,
United States.

MANUFACTURED LINENS

Damasks and Napkins

Scotland,
Ireland,
Germany,
Belgium.

Towelings

Scotland,
Ireland,
Germany,
United States,
Russia.

Glass Checks

Ireland.

Canvas

Scotland,
Ireland.

Handkerchief Lawns, Cambrics, and Laces

Ireland,
Germany,

France.

Towels

Germany,
Scotland,
Ireland,
Austria,
U. S. (union).

Linen Sheetings

Ireland,
Belgium,
France,
Scotland.

Blouse or Dress Linens

Ireland,
Scotland.

Bleached Waist Linens

Ireland,
France,
Belgium.

Fancy Linens, Doylies, etc.

Germany,
France,
Japan,
Madeira Islands,
Island of Teneriffe.

[18] The stem fibers such as flax, jute, ramie are called bast fibers, and before any of them can be utilized industrially, steps have to be taken to render them free from gum. When the stems of these plants are severed, the juice tends to oxidize through contact with the air and forms a gum of a peculiarly tenacious character.

[Pg 199]

CHAPTER XVI

HEMP

LOADING HEMP IN MANILA

Hemp is a fiber that is obtained from the hemp plant. It grows principally in Russia, Poland, France, Italy, Asia, India, the Philippines, Japan, and some parts of the United States—Kentucky, Missouri, Tennessee, Ohio, Indiana, and New York. The original country of the hemp plant was doubtless Asia, probably that [Pg 200] part near the Caspian Sea. The preparatory treatment is similar to that for the flax plant, except that most of the work is done by machinery. Considered chemically, in addition to cellulose, hemp fiber contains a considerable amount of woody matter, differing in this respect from cotton. Its properties are color (pearl gray, with green or yellow tints), fineness (which depends upon the quality of the hemp; it is usually bought as fine as flax), and tensile strength (which is considerable and greater than that of flax). Its best qualities are its slight luster and its ability to resist to a great extent the tendency to rot under water. Owing to the fact that it is difficult to bleach, it is used chiefly in making string, cord, ropes, etc.

Sisal Hemp. Sisal hemp is a variety that grows extensively in Central America and the West Indies. The plant, the *agava rigida*, is similar to what is known in this country as the century plant. The fiber is found in the leaves which closely surround the stalks. The common hemp on the other hand is found closely surrounding the woody part of the stem. The fiber of Sisal hemp is obtained by scraping away the fleshy part of the leaves with large wooden knives or by machines.

Manila Hemp. Manila hemp is obtained in the Philippines. The plant belongs to the banana family and grows as large as a small tree. The hemp is obtained from the leaf stalks which appear to form the trunk of the tree. The fiber is larger, not so stiff, but [Pg 201] stronger than Sisal hemp. The fiber of Russian hemp is the strongest; that of Italian hemp the finest.

FIELD OF SISAL HEMP

Jute. Jute is the name given to the fibers found in certain plants which grow principally in India, and the East Indian Islands. The common jute comes principally from the province of Bengal, India, where it was first known to science in 1725. The term jute was first applied to the fiber by Dr. Rosburgh in 1795. The plant is cut just about the time when it appears in full flower. The stalks are then bundled and retted by steeping in pools of stagnant water.

Jute occupies third position in importance of vegetable fibers in the manufacturing scale, being inferior to cotton and flax. Hemp is stronger than jute. Jute becomes weak when exposed to dampness.

It is extensively used for mixing with silk, cotton, flax, hemp, and woolen fabrics. The coarse varieties [Pg 202] are made into coarse fabrics—sacks, packing cloth, etc., while the finer varieties, in which the undesirable quality of growing darker with age is less apparent, are used for making carpets, curtains, and heavy plushes, for which they are very suitable.

[Pg 203]

CHAPTER XVII

SILK

Silk. The silk of commerce is obtained from the cocoons of several species of insects. These insects resemble strongly the ordinary caterpillars. At a certain period of its existence the silkworm gives off a secretion of jelly-like substance. This hardens on exposure to the air as the worm forces it out and winds it about its body.

MOTH, SILKWORM, AND COCOONS

It takes about three days for the worm to form the cocoon. After the cocoon has been formed the silkworm passes from the form of a caterpillar into a moth which cuts an opening through the cocoon and flies away. It is very important that the moth should not be allowed to escape from the cocoon; the mere [Pg 204] breaking of the cocoon greatly decreases the value of the thread. The cocoon is preserved by killing the chrysalis by heat.

There are a great many varieties of caterpillars, but few of them secrete a sufficient quantity of silk to render them of commercial value. The principal species is the mulberry silkworm which produces most of the silk in commerce. It is cultivated and fed on mulberry leaves. There are other varieties of silkworms that are not capable of being cultivated and are called wild silkworms. The silk produced by the wild worms of China and India is called "tussah" (or "tussur"). The silk is inferior to that produced by the cultivated worms and is used for making pile fabrics, such as velvet, plush, etc.

The color of the cocoons varies greatly. Most of the European cocoons are bright yellow, though some are white. The Eastern cocoons, on the other hand, are mostly white, while a few are yellow. The wild silks are for the most part écru color, though some are pale green. The color, except in the wild silks, is derived from the gum which is secreted by the worm, and with which the fibers are stuck together. This gum comprises from 15 to 30 per cent of the weight and is removed by boiling in soap and water before the silk is dyed. All silks except the wild silks, after the gum is removed, are from white to cream in color. The tussah, or wild silks, remain an écru color.

REELING RAW SILK

The greatest care has to be exercised throughout in the care of the moths, eggs, worms, and cocoons—this [Pg 205] being the succession of changes. That is, the moth lays eggs which are collected and kept cool till the proper season for incubation. They are then kept warm during the time occupied in hatching, sometimes about the person of the raiser. After a time these eggs hatch out worms, tiny things hardly larger than the head of a pin. After the worms are hatched they require constant care and feeding with chopped mulberry leaves till they reach maturity. They are then about three inches in length, and spin their cocoons from a fiber and gum which they secrete. When the cocoons are spun the worms become chrysalises inside of them. The cocoons are then collected and the chrysalises [Pg 206] killed, generally by heat, before they can again become moths.

Raw Silk. The cocoons are next sent to the reelers or filatures. A number of cocoons, greater or less, according to the size of thread desired, are placed in a basin of hot water, which softens the gum. After the outside fibers are removed so that the ends run free, the ends are collected through a guide and are wound upon a reel. As the silk cools and dries, the gum hardens, sticking the fibers from the different cocoons together in one smooth thread varying in size according to the number of cocoons used. After the silk has been reeled and dried it is twisted into hanks and sent to America and other countries as raw silk.

Most of the raw silk of commerce is produced in China, Japan, and Italy. It is also produced to a large extent in Italy, Turkey, and Greece, also France and Portugal. The cultivation of silk is not only carried on by private firms, but is encouraged by the government to the extent of granting money to the manufacturers.

Various attempts have been made to raise silkworms in the United States. All have failed on account of the high price of labor necessary to feed the worms.

Throwing. The manufacture in the United States begins with raw silk. We import our raw silk chiefly from Italy, China, and Japan. It is handled here first by the "throwster," who winds it from the skein and makes various kinds of thread for different purposes.

Raw silk wound on spools in a single thread, and [Pg 207] called singles is often used to make warps (that is, the threads running lengthwise of a piece of cloth) for piece-dyed goods, or cloth which is woven with the gum in the silk, and afterward boiled out and dyed. Singles are also sometimes used for filling (that is, cross threads) in very thin fabrics.

Silk yarn that is used for weaving is divided into two kinds, "tram" and "organzine." Tram silk is made by twisting two or more loosely twisted threads. It is heavier than organzine and is used for filling. Organzine silk is produced by uniting a number of strongly twisted threads. It is used for warp. Crêpe yarn is used in making crêpe, chiffon, and for other purposes. It is very hard twisted thread, generally tram, from forty to eighty turns per inch.

Embroidery silk is made by winding the raw silk, putting a large number of ends together, giving them a slack twist, then doubling and twisting in the reverse direction with a slack twist.

Sewing silk is made by winding and doubling the raw product, then twisting into tram, giving it a slack twist, doubling and twisting in the reverse direction under tension. Machine twist is similar, but three ply.

The principal fabrics made of silk are: silk, satin, plush, chenille, crêpe, crepon, gauze, damask, brocade, pongee, and ribbons. Silk thread and cord are also extensively used. The United States is among the leaders in the manufacture of silk fabrics.

Silk Waste. When the cocoons are softened for reeling a certain portion of the silk is found to consist of [Pg 208] waste and broken threads. The tangled silk on the outside of the cocoon is called floss. The residue after reeling, and other wastes in reeling, are known as frisonnets. Floss silk is not used for weaving. It is a slack twisted tram, generally composed of a large number of threads of singles.

Spun Silk. There is another class of threads made from waste silk by spinning and known as spun silk. Waste silks include the pierced cocoons, that is, those from which the moth has come out by making the hole and breaking the fibers in one end of the cocoon; the waste made in the filatures in producing raw or reeled silk, chiefly the outside fiber of the cocoon and the inside next the chrysalis; and also the waste made in manufacture. The waste silk is ungummed; that is, the gum is removed from the fibers by boiling with soap, by macerating or retting, or by chemical reagents.

After the gum is removed from the cocoons, they are opened and combed, most of the chrysalis shell being removed. The remainder, with other foreign matter, is picked out by hand from the combed silk. The silk is put through a number of drawing frames to get the fibers even on the roving frames, where it first takes the form of thread, then on the spinning frames, where it is twisted. If it is to be used as singles, the manufacture ends here. In two- or three-ply yarns, the singles are doubled, twisted again, singed by running through a gas flame, cleaned by friction, controlled, that is, the knots and lumps taken out, and then reeled into skeins for dyeing or put on spools.

[Pg 209] **Spun Numbers.** There are two methods in general use for numbering spun silk. In the French system, the number is based on the singles, by the meters per kilogram; two and three cord yarns have one-half, one-third, etc., the length the numbers indicate. Thus—

No. 100 singles has 100,000 meters per kilogram.
No. 2-100 has 50,000 meters per kilogram.
No. 3-100 has 33,333 meters per kilogram.

The other system which is more generally used in this country, is the English system. The hank is 840 yards, and the number of hanks in one pound avoirdupois is the count of the yarn. It is based on the finished yarn, and singles, two or three cord yarns of the same number all have the same yards per pound. Thus—

No. 50 singles has 42,000 yards per pound.
No. 50-2 has 42,000 yards per pound.
No. 50-3 has 42,000 yards per pound.

Dyeing Yarns. Generally speaking there are two large classes into which silk goods may be divided, those in which the threads are colored before weaving and called yarn-dyed goods, and those dyed or printed after weaving and called piece-dyed or printed goods. In dyeing yarns, the silk is first ungummed and cleaned by boiling in soap and water, then washed in cold water. If the thread is to be weighted, as is frequently done, tin salts, iron, or other heavy material is deposited on the fiber. If carried far, this is injurious, making the silk tender and weak. Sometimes there is more weighting than silk. Yarns are usually dyed in hot [Pg 210] liquors, aniline colors being the ones in most common use to-day, though other dyes are used for special purposes. Some yarns are dyed in the gum, and some with a part of the gum left in. After dyeing, they are washed in cold water, dried, and wound on spools.

Silk Dyeing. Silk occupies in several respects an intermediate position between the animal and vegetable fibers. Like wool, it is a highly nitrogenous body, but contains no sulphur. It readily takes up many of the colors which can be worked upon vegetable fiber by the aid of the mordants. This is particularly the case with reference to a large number of aniline colors, which require merely to be dissolved and mixed with perfectly clear water in the dye vessel. The great attraction of silk for these colors simplifies silk dyeing exceedingly. The sad colors, on the other hand, and especially black, are in many cases exceedingly complex, the main object of the dyer being not so much to color the silk as to increase its weight.

Dyeing black on silk is unquestionably the most important branch of silk dyeing, and it has probably received more attention than any other branch, in consequence of which it has been brought to a high degree of perfection. Blacks on silks are produced both from natural and artificial coloring matters, the former having, so far, retained their pre-eminence despite the recent discoveries of chemists. For various reasons coal-tar colors have never proved successful in dyeing black on silk. Since the discovery of America, logwood blacks have formed the staple of the black-silk dyer, [Pg 211] who has carried their production to a high degree of excellence. But unfortunately, besides aiming at a high state of perfection in the actual dyeing operation, the black-silk dyer has also aimed at increasing the weight of the dyed silk, so that nowadays it is possible for him to receive ten pounds of raw silk and to send out fifty pounds of black silk, the extra forty pounds being additions made in the process of dyeing.

WINDING SILK ON SWIFTS

Logwood black-silk dyeing consists essentially of alternate dippings in separate baths with the mordant and dyestuffs suitable for producing the required color and weight. The number of dippings and the length [Pg 212] of time taken in each operation depend on the intensity of the black wanted and the amount of weighting which is desired. The chief substances used for weighting are lead salts, catechu, iron, and nut-galls, with soap and oil to soften in some degree the harshness of the fabric which these minerals cause. As the details of the operations are practically the same for all kinds of logwood blacks (raven, jet, crape, dead black, etc.), the method for producing one will suffice for all. The process involves several distinct operations, as follows:

1. **The Boiling Off.** This is the removal of the gum and natural coloring matter in the silk. It is accomplished by boiling the skeins of silk in water and good olive oil soap for about one hour. This dissolves the gum and leaves the fiber clean and glossy.

2. **Mordanting.** This is done in a bath of nitrate of iron, in which the skeins of silk are allowed to remain one hour. The silk gains some in weight in this operation by absorbing a quantity of the iron in the bath. After having been dipped in the first bath three or four times, it is ready for the soap and iron bath, in which it is repeatedly immersed, the operation causing a deposit of iron-soap on the fiber which adds to its weight, but at the same time does not lessen its flexibility and softness. Eight dippings in the iron and soap bath increase the weight of the silk about 100 per cent.

3. **Blue Bottoming.** The next operation is to dye the silk blue, which is done by immersing it in a solution of potash. In this it is worked for half an hour, when it acquires a deep blue color. It is then taken [Pg 213] out, and after rinsing is ready for the "weighting" operations.

4. **"Weighting" Bath.** A catechu bath is now prepared, in which the silk is entered and worked for an hour, and then allowed to steep over night. The result is that the blue on the silk is decomposed, and the goods by absorbing the tannin in the catechu increase in weight from 35 to 40 per cent. This bath is the most important one in the dyeing of "weighted" black silks, as the dyer can regulate the strength of the bath by the addition of tin crystals so as to increase the weight of the silk to an astonishing degree. The proportion of tin crystals used is regulated by the number of iron baths that have previously been given the silk; if two baths of iron have been given, 5 per cent of tin crystals are used; if four baths, 10 per cent, and so on. The action of these chemicals is somewhat complex. All that is known is that by reason of some peculiar quality possessed by silk it is enabled to combine with iron and tin, and that exposure to the air after the baths fixes these chemicals permanently upon the fibers, thus increasing their weight to almost any desired extent. Silk, according to its quality and weight, will take up of these substances from 50 to 200 per cent without creating much suspicion. Instances have been known in which silk has been increased nine times its own weight. All the operations thus far have had for their object the weighting of the silk, although the blueing and the catechu baths have some influence on the finished result. After these come the dyeing [Pg 214] operations proper, two in number, mordanting and dyeing.

5. **Mordanting.** A bath of iron liquor heated to 130 degrees F. is provided. The silk is entered, worked well for one hour, then wrung out and hung up to "age" for two hours, after which it is ready for the logwood dye.

6. **Dyeing.** A bath of logwood liquor is prepared to which is added 10 per cent of fustic, and the solution is brought to a temperature of 150 degrees F. In this the silk is entered and worked for an hour, then taken out and wrung dry. Sometimes the black does not come up full enough, and in such cases the bath is repeated.

7. The final operation has for its object the restoration of the luster and suppleness of the silk, which has to some extent deteriorated from the many operations through which it has passed. The brightening and softening of the fiber are effected by immersing the silk in a bath of olive oil in the form of an emulsion. In this the silk is worked until it is thoroughly impregnated with the oil, when it is taken out and wrung dry, after which it is ready for the loom. Practically the same process is followed in piece dyeing, though only inferior grades of silk are dyed in the web.

Colored Silks. This class of silks is generally purer than black and sad-colored silks. It is not nearly so easy to weight the former as the latter, for the reason that there are but few substances capable of giving weight which do not interfere with the effect of light colored dyes. The weighting agents most generally [Pg 215] used are sugar and acetate of lead. The weighting by sugar is done after the silk is dyed. A solution is made of pure lump sugar by placing it in a large copper pan with water and heating until dissolved. In this bath the silk is thoroughly saturated, and then dried and finished; or, the dipping process may be repeated several times if desired. One dipping will weight the silk about 12 per cent, two about 20 per cent, and three about 30 per cent. In a solution of acetate of lead, each dipping will weight the silk about 8 per cent, and these may be repeated as often as it is wished. In this case the weighting is generally done on the undyed, boiled-off silk, although it may be done on the dyed silk if the color is such as will stand the acid.

Mixed Silk Fabrics. Until lately silk was invariably dyed in the state of yarn. When the silk was to be woven into mixed fabrics, such as satin, gloria, etc., it was impossible to dye both fibers exactly the same shade. Formerly such fabrics were woven with the cotton and silk yarns dyed separately, care being taken to match them as closely as possible. The weaving of dyed yarns of different fibers is open to the objection that when the fabric comes to be finished there is a wide difference in the color, no matter how closely they may have matched in the beginning.

Ribbons. Ribbons are woven several pieces in one loom, with a separate shuttle for each piece. The shuttle is carried through the shed or warp by a rack and pinion, instead of being thrown through as in broad goods; otherwise the weaving is the same.

[Pg 216] **Velvets.** Velvets and other pile fabrics are woven in two pieces, one over the other, with the pile threads woven back and forth between them. A knife travels between the two pieces cutting the pile threads so as to leave the ends standing up straight. Velvets used to be woven over wires and cut by hand, but this method is practically obsolete.

Piece Dyeing. If the goods are woven with the gum still in the silk, it must be taken out afterward, and the goods either dyed in the piece or prepared for printing.

Printing. The most primitive method of printing is by the use of stencils. It is the method employed by the Japanese and Chinese. Next came block printing, which is still extensively used in Europe. The pattern is raised in felt on wooden blocks, the color taken up from pads, one block for each color. The results are good, but the work is very slow. Most silk goods are to-day machine printed. The design is engraved or etched on copper cylinders, one cylinder for each color; the color thickened with gum is supplied by rolls running against the cylinders, and the surplus is scraped off by a knife blade, leaving only that in the engraving which is taken up by the cloth. After printing, the cloth is steamed to set the colors, and then washed in order to remove the gum used to thicken the colors for printing.

[Pg 217]

JAC-
QUARD SILK LOOM

[Pg 218] **Finishing.** All silk goods, whether yarn dyed or piece dyed or printed, are given some kind of finish; sometimes it is no more than is necessary to smooth out the wrinkles. There are many finishing processes by which goods may be treated. They are run through gas flames to singe off loose fiber, and over steam cylinders to dry and straighten them, over a great variety of sizing machines to stiffen them with starch or glue. There are calenders or heavy rolls to smooth and iron them, steam presses of great power to press them out, breaking and rubbing machines to soften them, and tentering machines to stretch them to uniform width. There are also moireing or watering, embossing, and various other machines for special purposes.

Waterproofing. One of the worst difficulties with which the manufacturer of piece-dyed and printed silk goods has to contend is the ease with which they become spotted with water, and for a number of years many people have tried to prevent this by various processes. There are no less than two hundred such processes patented. None of them have met with much success, as they injured the feel or strength of the goods. After goods are finished they are carefully inspected for imperfections, measured, and wrapped in paper and packed in cases for shipment. The complexity and number of processes for treating silk goods may be realized when we know that a piece-dyed or printed fabric is handled its entire length between fifty and one hundred times after it comes from the loom, sometimes even more.

[Pg 219]

CHAPTER XVIII

PRINCIPAL SILK FABRICS

Alma. Cloth, double twilled from left to right diagonally, first made in black only as a mourning fabric. The name is from the Egyptian, as applied to a mourner or a singer at a funeral.

Barège. Sheer stuff of silk and wool for veiling, named from the town of Barèges, in France.

Bengaline. An imitation of an old silk fabric made for many centuries in Bengal, India, whence the name. The weave is similar to that of ordinary rep or poplin, being a simple round-corded effect. The cord is produced by using a heavy soft-spun woolen weft which is so closely covered by the silk warp threads that it is not exposed when examined from the wrong side. The same weave is also found in all-silk goods, under the designation of all-silk bengaline. When cheapened by the use of a cotton weft in place of wool the fabric is known as cotton bengaline, although the cotton is in the filling only.

Berber. Satin-faced fabric of light-weight cloth. It came into favor about the time of the defeat of the Berbers by General Gordon in his campaign against the Mahdi in North Africa.

Brocade. Raised figures on a plain ground.

[Pg 220] **Brocatel.** A kind of brocade used for draperies and upholstery; usually raised wool figures on a silk ground.

Bombazine. Silk warp, wool weft, fine twilled cloth; originally made in black only for mourning. It is used largely for mourning hat bands. The root of the name is *bombyx*, the Latin for silkworm.

Chenille. Cloth of a fuzzy or fluffy face; woven of cotton, silk, or wool; used sometimes for dress goods; more generally for curtains and table covers. *Chenille* is the French word for caterpillar, which the single thread of the cloth resembles.

Chiffon. A thin, transparent silk muslin. Although one of the thinnest and gauziest of modern silk fabrics, it is relatively strong considering its lightness. To convey an idea of the fineness of the thread used in its manufacture, it is stated that one pound of it will extend a distance of eight miles. In the process of finishing the fabric receives a dressing of pure "size." There are two styles of finish, called respectively the demi- or half size and the full size. Chiffon finished by full sizing is comparatively stiff; while the demi-finish produces a softer and lighter texture. It is dyed in a great variety of colors, and sometimes is printed in delicate patterns. It is especially adapted for home and evening wear, and is used for neck and sleeve trimming, drapery over silk foundations, fancy work, and millinery.

China Silk. A term applied to plain woven silks manufactured in China. The term China silk has been adopted in the United States in recent years for a class [Pg 221] of machine-woven silks made in imitation of the hand-loom product. These imitations are narrow in width and lack the soft, lustrous quality of Eastern fabrics, and are also free from the uneven threads. China silks are distinguished by their irregular threads, caused by some of the threads being heavier than others, and their extreme softness.

The warp and filling are identical in size and color, and being woven evenly produce a beautiful natural luster. It is generally plain color, although the figured goods are printed in much the same manner as calico. It is used for gowns, waists, underclothing, etc. It launders as well as white cotton.

Crêpe. A thin, gauzy fabric, woven in loose even threads of silk, heavily sized or gummed, crimped or crêped in the dyeing. Crêpe was first used in black only as a badge of mourning. It is now an accepted dress fabric, made in colors and white and of many materials. The name signifies to crimp or crêpe with a hot iron.

Crêpe de Chine. A soft, lustrous silk crêpe, the surface of which is smoother than that of the ordinary varieties. It is woven as a plain weave with part of the warp threads right twisted and the rest left twisted. It is dyed almost any color and figured or printed.

Eolienne. Sheer cloth of silk, silk and wool, or silk and cotton, woven in fine card effect. The name comes from the Greek Æolus, god of the winds.

Foulard. Plain silk cloth, sold as dress goods; originally [Pg 222] made for handkerchiefs only. The name is French for silk handkerchief.

Glacé. Plain, lustrous silk, yarn dyed, with warp of one color, and weft of another. The name is applied to all fabrics having two tones. Glacé is French for icy, having an icy appearance.

India Silk. A name applied to the plain woven silks manufactured in India on the primitive hand looms. The warp and weft are woven evenly and produce a beautiful natural luster. It is similar to China and Japanese silk. In fact most of these fabrics come from China and Japan, India silk being almost unknown in this country as so little of it is exported. The durability of these silks is about the same, and there is little difference in the prices.

Japanese Silk. A term applied to the plain woven silk manufactured in Japan. The warp and filling of this fabric are identical in size and color, and being woven evenly produce a beautiful natural luster. The weave is smooth and soft in quality. It is dyed in plain colors. The figured goods are printed in much the same way as calico. It is used for waists, gowns, and fancy underwear.

Jersey Cloth. Silk jersey cloth is popular at present. It is a knitted silk fabric, not woven, and is generally dyed in plain colors. It is expensive and is used for women's dresses, wraps, and silk gloves.

Meteor. Crêpe de meteor was originally a trade name for crêpe de chine, but now applied to a fabric which is distinguishable from crêpe de chine.

[Pg 223] **Moire.** Moire is a waved or watered effect produced upon the surface of various kinds of textile fabrics, especially on grosgrain silk and woolen moreen. This watered effect is produced by the use of engraved rollers and high pressure on carded material. The object of developing upon woven textiles the effect known as moire is the production of a peculiar luster resulting from the divergent reflection of the light rays on the material, a divergence brought about by compressing and flattening the warp and filling threads in places, and so producing a surface the different parts of which reflect the light differently. The moire effect may be obtained on silk, worsted, or cotton fabrics, though it is impossible to develop it on other than a grained or fine corded weave. The pressure applied to the material being uneven, the grained surface is flattened in the parts desired. In the Middle Ages moire was held in high esteem, and continues to enjoy that distinction down to the present day. It is used for women's dresses, capes, and for facings, trimmings, etc.

Mozambique. Grenadines, with large colored flower designs in relief.

Organzine. Silk fabric, made with warp and filling of the same size. Organzine is the name given the twisted silk thread in Italy, where it is made.

Panne. This name is applied to a range of satin-faced velvet or silk fabrics which show a high luster produced by pressure. The word *panne* is the French for plush.

Peau de Soie. Literally, skin of silk. A variety of [Pg 224] heavy, soft-finished, plain-colored dress silk, woven with a pattern of fine close ribs extending weftwise of the fabric. An eight-shaft satin with one point added to the original spots on the right or left, imparting to the fabric a somewhat grainy appearance. The best grades of peau de soie present the same appearance on both sides, being reversible. The lower grades are finished on one side only.

Plush. Long piled fabric of the velvet order. *Peluché*, the origin of the name, is French for shaggy.

Pongee. Said to be a corruption of Chinese *punchi*, signifying home made or home woven. Another suggestion is that the word is a corruption of *pun-shih*, a native or wild silk. A soft, unbleached, washable silk, woven from the cocoons of the wild silkworm, which feeds on the leaves of the scrub oak. Immense quantities in a raw state are annually shipped from China to this country and Europe, where they are bleached, dyed, and ornamented with various styles of designs. The name is also applied to a variety of dress goods woven with a wild silk warp and a fine worsted weft.

Popeline. A French name. The French fabric is said to have been first introduced during the early part of the sixteenth century at Avignon, then a papal diocese, and to have been so called in compliment to the reigning pope. A fabric constructed with a silk warp and a filling of wool heavier than the silk which gives it a corded surface. Poplin manufacture was introduced into Ireland in 1693 by a colony of fugitive French Huguenots. The industry concentrated at Dublin, [Pg 225] where it has since remained. The Irish product has been celebrated for its uniformly fine quality. It is always woven on hand looms, which accounts for the high price it commands in English and American markets. The wool used is a fine grade of Cape or Australian, which is the most suitable in texture and length of fiber. The silk is unweighted Chinese organzine. The result is a rich, handsome fabric resembling whole silk goods in appearance, but inferior to them in durability and produced at a much less cost. It is used for ladies' waists, wraps, and gowns.

Figured Poplin. A stout variety, ornamented in the loom with figures. The ground is composed of clear, sharp cords extending across the web. It is sometimes woven entirely of silk, but oftener of silk and wool. Used for high-class upholstery purposes, and for curtains and hangings.

Terry Poplin. A silk and wool dress fabric in the construction of which the alternate warps are thrown upon the surface in the form of minute loops.

Sarsenet. A thin, soft-finished silk fabric of a veiling kind, now used as millinery lining. The name comes from the Arab Saracens, who wore it in their head-dress.

Satin. When satin first appeared in trade in Southern Europe it was called *aceytuin*. The term slipped through early Italian lips into *zetain*, and coming westward the *i* was dropped, and smoothed itself into satin. There is evidence that the material was known as early as the fourteenth century in England, and probably in [Pg 226] France and Spain previous to that time, though under other names.

In the weaving of most silk fabrics the warp and filling intersect each other every alternate time (as in plain weaving), or every third or fourth time (as in ordinary twill weaving) in regular order; but in weaving satin the fine silk warp only appears upon the surface, the filling being effectually covered up and hidden. Instead of making the warp pass under and over the filling every alternate time, or over two or three filling threads in regular order, it is made to pass over eight, ten, twelve or more filling threads; then under one and over eight more, and so on. In passing over the filling, however, the warps do not interweave at regular intervals, which would produce a twill, but at irregular intervals, thus producing an even, close, smooth surface, and one capable of reflecting the light to the best advantage. The filling of low grade satin is generally cotton, while in the better goods it is silk. Common satin is what is technically known as an eight-leaf twill, the order in which the filling thread rises being once in eight times. Rich satins may consist of sixteen-leaf to twenty-leaf twills. The cheap qualities of cotton-back satin, particularly those that sell at wholesale for fifty cents and under, are not made to any extent in this country, our manufacturers being unable to compete with foreign mills in these lines.

Satins are woven with the face downward, because in weaving, say a sixteen-leaf satin, it would be necessary, were the surface upward, to keep fifteen heddles [Pg 227] raised and one down, whereas with the face of the cloth under, only one heddle has to be raised at a time. When first taken from the loom the face of satin is somewhat flossy and rough, and hence requires to be dressed. This operation consists of passing the pieces over heated metal cylinders which remove the minute fibrous ends, and also increase the natural brilliance of the silk. Cotton-back satins are used by coffin manufacturers, fancy box makers, fan makers, and by the cutting-up trade. Rich satins are used in making ladies' gowns and waists.

Soleil. Satin-faced cloth, woven with a fine line, a stripe running lengthwise of the piece. It is usually made in solid colors and piece dyed. *Soleil* is French for sun, and applies to the brightness of the finished cloth.

Taffeta. Derived from Persian *taftah*. Taffeta is one of the oldest weaves known, silk under this name having been in constant use since the fourteenth century. During this long period the term has been applied at different times to different materials. It is a thin, glossy silk of plain texture or woven in lines so fine as to appear plain woven. The weave is capable of many effects in the way of shot and changeable arrangements, which are produced by threads of different colors rather than by any special disposition of warp and filling. Taffeta has the same appearance on both sides. It is piece dyed in numberless plain colors, and also produced in a great variety of ornamental patterns, such as fancy plaids, cords, and stripes (both printed and woven). [Pg 228] The following considerations contribute chiefly to the perfection of taffetas, viz.: the silk, the water, and the fire. The silk must not only be of the finest kind, but it must be worked a long time before it is used. The watering, which is given lightly by any acidulous fluid, is intended to produce the fine luster, and lastly, the fire and pressure which have a particular manner of application. Its wearing qualities are not of the best. The cloth cracks or breaks, especially if plaited. It is used for gowns, shirtwaists, linings, petticoats, etc.

Tulle. Openwork silk net; made on the pillow as lace by young women of Tulle, France.

Velour. French for velvet. A trade term of somewhat loose application, being used indiscriminately to describe a great variety of textures so constructed or finished as to present a velvet-like surface. It is usually a velvety fabric made of coarse wool yarn and silk. Velour is woven with a coarse stiff pile after the manner of plush; while at present it is made of jute, cotton, and worsted, it was originally constructed of linen. It is produced in numberless forms, both plain and in fancy effects.

Velvet. From the Italian *velluto*, feeling woolly to the touch, as a woolly pelt or hide. Fine velvet is made wholly of silk.

Velveteen. An imitation velvet, made of cotton, usually with plain back, not twilled, as silk velvet.

Tabby Velvet. The lowest grade of cotton velvet, used for covering cheap coffin lining cases, sold by the inch in widths which range from sixteen to thirty-two [Pg 229] inches. Originally made in Bagdad for wall covering, its name being derived from a section of that city.

Voile. From the French *voile*, meaning a veil, a light fabric usually more or less transparent, intended to conceal the features in whole or in part or to serve as a screen against sunlight, dust, insects, etc., or to emphasize or preserve the beauty. The custom of wearing veils had its origin in the early ages in the desire of semi-savage man to hide away the woman of his choice, and is a survival of the ancient custom of hiding women that is found even down to the present day in Eastern countries. Voile is a transparent, wiry material with a square mesh.

[Pg 230]

CHAPTER XIX

ARTIFICIAL SILK

Silk Cotton. On account of the high price of silk various attempts have been made to find satisfactory substitutes for it. There are certain seed coverings of plants that contain very fine hair-like fibers with a luster almost equal to silk, but the staples are short, and the texture weak. The Kapok plant furnishes most of the commercial silk cotton on the market. The fibers of Kapok are thin and transparent. They are extremely light, and the length is less than half an inch. Silk cotton has a smooth surface and therefore cannot be spun like true cotton which has corded edges.

Artificial Silk. Since seed hairs are composed, like all vegetable fibers, of cellulose, attempts have been made to prepare an artificial silk product from waste paper—that is, by treating waste paper or wood or cotton fibers with various chemicals in order to obtain pure cellulose. This artificial silk is perhaps the most interesting of artificial fibers, but its manufacture is dangerous, owing to the ease with which it catches fire and explodes. Cellulose, chemically treated, can be transformed into a fluid solution known as collodion. The collodion is placed in steel cylinders and expelled by pressure through capillary tubes. After drying, [Pg 231] denitration, and washing, it may be spun and dyed like natural silk. Colored threads may be produced by the addition of certain dyes.

Artificial silk bears a deceptive resemblance to the natural article, and has nearly the same luster. It lacks the tensile strength and elasticity, and is of higher specific gravity than true silk.

Tests. A simple way of recognizing artificial silk is by testing the threads under moisture, as follows: First, unravel a few threads of the suspected fabric, place them in the mouth and masticate them vigorously. Artificial silk readily softens under this operation and breaks up into minute particles, and when pulled between the fingers shows no thread, but merely a mass of cellulose or pulp. Natural silk, no matter how thoroughly masticated, will retain its fibrous strength. The artificial silk offers no resistance to the teeth, which readily go through it; whereas natural silk resists the action of the teeth.

[Pg 232]

CHAPTER XX

SUBSTITUTES FOR COTTON

On account of the high price of cotton various experiments have been made in an effort to replace it with fiber from wood pulp, grasses, leaves, and other plants.

Wood Pulp. A Frenchman has discovered a process, *la soyeuse*, of making spruce wood pulp into a substitute for cotton. Although it is called a substitute, the samples show that it takes dye, bleaching, and finishing more brilliantly than the cotton fiber. It resists boiling in water or caustic potash solution for some minutes, and does not burn more quickly than cotton. The fiber can be made of any length, as is also the case with artificial silk. The strength of the yarn apparently exceeds cotton, and the cost of manufacture is much lower. Arrangements are being made in Europe for the extensive production of this fiber.

Ramie. Ramie or China grass is a soft, silky, and extremely strong fiber. It grows in southwestern Asia, is cultivated commercially in China, Formosa, and Japan, and is a fiber of increasing importance. Ramie is a member of the nettle family and attains a height of from four to eight feet. After the stalks are cleaned of a gummy substance, insoluble in water, it is known as China grass, and is used in China for summer clothing. [Pg 233] In Europe and America by the use of modern machinery and chemical processes the fiber is cleaned effectively and cheaply. After it is bleached and combed it makes a fine silky fiber, one-half the weight of linen, and three times stronger than hemp. It is used in Europe to make fabrics that resemble silk, and is also used in making underwear and velvets. With other fabrics it is employed as a filling for woolen warps. It will probably be used widely in the United States as soon as cheaper methods of cleaning are devised.

Pineapple and Other Fibers. Other fibers, of which that from the pineapple is the most important, are used for textile purposes in China, South America, parts of Africa, Mexico, and Central America. Their use has not been extensive on account of high cost of production. The silk from the pineapple is very light and of excellent quality.

Spun Glass. When a glass rod is heated in a flame until perfectly soft it can be drawn out in the form of very fine threads which may be used in the production of handsome silky fabrics. Spun glass can be produced in colors; but on account of the low elasticity of these products, their practical value is small, though the threads are exceedingly uniform and have beautiful luster. Spun glass is used by chemists for filtering strong acid solutions.

A kind of glass wool is produced by drawing out to a capillary thread two glass rods of different degrees of hardness. On cooling they curl up, in consequence of the different construction of the two constituent threads.

[Pg 234] **Metallic Threads.** Metallic threads have always been used for decorating, particularly in rich fabrics. Fine golden threads, as well as silver gilt threads, and silver threads and copper wire, have been used in many of the so-called Cyprian gold thread fabrics, so renowned for their beauty and permanence in the Middle Ages. These threads are now produced by covering flax or hemp threads with a gilt of fine texture.

Slag Wool. Slag wool is obtained by allowing molten slag (generally from iron) to run into a pan fitted with a steam injector which blows the slag into fibers. The fibers are cooled by running them through water, and the finished product is used as a packing material.

Asbestos. Asbestos is a silicate of magnesium and lime, containing in addition iron and aluminum. It is found in Savoy, the Pyrenees, Northern Italy, Canada, and some parts of the United States. Asbestos usually occurs in white or greenish glassy fibers, sometimes combined in a compact mass, and sometimes easily separable, elastic, and flexible. Canadian asbestos is almost pure white, and has long fibers. Asbestos can be spun into fine thread and woven into rope or yarn, but as it is difficult to spin these fibers alone, they are generally mixed with a little cotton, which is afterwards disposed of by heating the finished fabric to incandescence. Because of its incombustible nature asbestos is used where high temperatures are necessary, as in the packing of steam joints, steam cylinders, hot parts of machines, and for fire curtains in theatres, hotels, etc. It is difficult to dye.

[Pg 235]

APPENDIX

Testing Textile Fabrics. This is an age of adulteration, and next to food there is probably no commodity that is adulterated as much as the clothing we wear. Large purchasers of textile fabrics and various administrative bodies, such as army clothing departments, railway companies, etc., have adopted definite specifications to ensure having good material and workmanship. Before the fabrics are accepted they are examined carefully by certain tests to see if they meet the requirements. Wholesale and retail merchants insist on various conditions when purchasing fabrics in order to conform to the increasing needs of the public. Hence every manufacturer, buyer, or dealer in fabrics should be familiar with the tests used to determine the quality of goods he is about to buy.

The tests used are as follows:

1. Identification of the style of weaving.

2. Testing the breaking strength and the elasticity by the dynamometer.

3. Determining the "count" of warp and filling.

4. Determining the shrinkage.

5. Testing the constituents of warp and of filling.

6. Testing the finish and dressing materials.

7. Testing the fastness of the dye.

[Pg 236] **Directions for Determining the Style of Weave.** In examining a fabric for the weave it is first necessary to determine the direction of the warp and filling threads. This is a very simple matter in a great many fabrics that have a selvedge—the warp must be parallel to the selvedge.

In fabrics that have been fulled, raised, and cropped, as buckskin, flannel, etc., the direction of the nap will indicate the direction of the warp, since the nap runs in this direction.

In the case of fabrics with doubled and single threads, the doubled threads are always found in the warp.

In fabrics composed of cotton and woolen threads running in different directions, the cotton yarn usually forms the warp and the woolen yarn the filling. Then again the warp threads of all fabrics are more tightly twisted than the filling threads, and are separated at more regular intervals.

Sometimes in stiffened or starched goods threads running in only one direction can be seen. In this case they are the warp threads.

If one set of threads appears stiffer and straighter than the other, the former may be regarded as warp, while the rough and crooked threads are the filling. The yarn also gives one a hint, since the better, longer, and higher number material constitutes the warp, while the thicker yarn the filling.

The direction of the twist of the thread is conclusive; if one set has a strong right twist and the other a left twist the first is the warp.

[Pg 237] After determining the direction of the warp and filling, the next point is to determine the interlacing of the warp and filling threads—the weave. This may be done by inspection or by means of a pick-glass and needle. The weave may be plotted on design paper (plotting paper), the projecting warp threads being indicated by filling up the corresponding square, and leaving those referring to the filling threads blank. In this way the weaving pattern of the sample is obtained, and serves as a guide to the weaver in making the fabric, as well as for the preparation of the pattern cards for the Jacquard loom.

Testing the Strength and Elasticity of a Fabric. The old-fashioned plan of testing cloth by tearing it by the hand is unreliable, because tearing frequently requires only a certain skilled knack whereby the best material can be pulled in two. In this way an experienced man may tell good from bad cloth, but he cannot determine slight differences in quality, because he has exerted his strength so often that his capacity to distinguish the actual force has disappeared.

The best means of determining the strength of a fabric is by means of a mechanical dynamometer, [19] which expresses the tensile strength of the fabric in terms of weight. The machine is very useful to the manufacturer because it enables him to compare accurately his various products with those of his competitors. The value of these tests is sufficiently proved by the fact that all [Pg 238] army clothing departments, etc., require their supplies of cloth, etc., to pass a definite test for strength.

Breaking tests also afford the most certain proof to bleachers of cotton and linen goods as to whether the bleaching has burned or weakened the goods. The same test will quickly determine whether a fabric has been improperly treated in the laundry.

Determining the Count of Warp and Filling Threads. Every fabric must contain a certain count of warp and filling threads—a definite number within a certain space for each strength of yarn employed. A fabric is not up to the standard of density when less than the requisite number of warp or filling threads per inch is found. For example, if a buyer was told that a fabric is 80 square, that is, eighty warp threads and eighty filling threads to the inch, and on examination found only 72 square, he would immediately reject the goods.

The count of warp and filling is determined by means of a pick-glass—a small mounted magnifying glass—the base of which contains an opening of one-half inch by one quarter inch, or one quarter inch by one quarter inch. If the pick-glass is placed on the fabric the number of warp and filling threads may be counted, and the result multiplied by either two or four, so as to give the number of threads to the inch. For example, if I count twenty picks and twenty threads on a one quarter-inch edge, there are eighty picks and eighty threads to the inch. A more accurate result can be obtained by using a pick-glass with a one-inch opening.

[Pg 239] **Determination of Shrinkage.** A very important factor in the value of a fabric is the shrinkage. The extent of this may be determined by pouring hot water over a sample of about twelve by twenty inches, and leaving the fabric immersed over night, then drying it at a moderate temperature without stretching. The difference in length gives the shrinkage, which is usually expressed in percentage.

Determination of Weight. Buyers and sellers of dry goods, when traveling, are anxious to determine the weight of fabrics they examine. This may be done by means of small pocket balances so constructed as to give the number of ounces to the yard of a fabric.

Testing the Constituents of the Warp and Filling. Take a sample piece of the cloth to be examined—the piece must be large enough to contain specimens of all the different kinds of yarn present in the material—and separate all the filling and warp threads. Be sure that all double threads are untwisted.

Combustion Test; Test for Vegetable and Animal Fibers. Burn separately a sample of the untwisted warp and filling threads. If one or both burn quickly without a greasy odor, they are vegetable fibers, cotton or linen. If one or both burn slowly and give off a greasy odor, they are animal fibers, wool or silk. This test is not conclusive, and further chemical examination—acid test—must be made to ascertain whether wool is pure or mixed with cotton.

Acid Test. The vegetable fibers, cotton and linen, are distinguished from those of animal origin by their [Pg 240] behavior in the presence of acids and alkalies. The vegetable are insoluble when boiled with a 4 per cent sodium hydrate solution, but readily clear or carbonize when saturated with a 3 per cent sulphuric acid solution and allowed to dry at a high temperature in a hot closet. Wool on the other hand is not affected by the action of weak sulphuric acid.

Cotton Distinguished from Linen. If the fibers are vegetable, cotton may be distinguished from linen by staining the fibers with fuchsine. If the fibers turn red, and this coloration disappears on the addition of ammonia, they are cotton, if the red color remains the fibers are linen. Whenever cotton yarn is used to adulterate other fabrics, it wears shabby and loses its brightness. When it is used to adulterate linen, it becomes fuzzy through wear. One may detect it in linen by rolling the goods between thumb and finger. Linen is a heavier fabric, and wrinkles much more readily than cotton. It wears better, and has an exquisite freshness that is not noticed in cotton fabrics.

Silk Distinguished from Wool. Place the fabric or threads containing animal fibers in cold, concentrated hydrochloric acid. If silk is present it will dissolve, while wool merely swells.

208

Artificial Silk from Silk. On account of the low value of the artificial and the high value of genuine silk, there is a tendency to offer the artificial instead of the pure article. Test: When artificial silk is boiled in 4 per cent potassium hydrate solution it produces a yellow solution, while pure silk gives a colorless solution.

[Pg 241] A common test is to put the artificial silk in water, where it will pull apart as though rotten; or to take out one strand of the silk, hold it between the finger and thumb of each hand and wet the middle of the strand with the tongue, when it will pull apart as though rotten.

Artificial silk is inferior in strength and elasticity to pure silk. Then again it is lacking in the crackling feeling noticed in handling the genuine article.

Test for Shoddy. It is no easy matter to detect shoddy in woolen fabrics; the color of the shoddy threads is the best evidence. Many parcels of rags are of one single color, but for the most part they are made of various colored wools; therefore, if on examination of a fabric with a magnifying glass a yarn of any particular color is found to contain a number of individual fibers of glaring colors, the presence of shoddy can be assumed with certainty.

Woolen goods containing cotton are seldom made from natural wool. Shoddy yarns, especially in winter goods, are found in the under-filling at the reverse side of the cloth, as thick, tightly twisted yarns, curlier than those from the pure wool.

Determination of the Dressing. During the various operations of washing, bleaching, etc., the goods lose in weight, and to make up this deficit a moderate amount of dressing or loading is employed. Dressing is not regarded as an adulteration, but as an embellishment.

Various dressing materials are used, such as starch, flour, mineral matters, to give the goods stiffness and feel on one hand, and on the other to conceal defects [Pg 242] in the cloth, and to give a solid appearance to goods of open texture. The mineral substances used serve chiefly for filling and weighting, and necessitate the employment of a certain quantity of starch, etc. In order that the latter may not render the cloth too stiff and hard, further additions of some emollient, such as glycerine, oils, etc., are necessary.

When a fabric filled in this manner is placed in water and rubbed between the hands, the dressing is removed, and the quantity employed can be easily determined.

By holding fabrics before the light dressing will be recognized, and such goods, if rubbed between the fingers, will lose their stiffness. Loading is revealed by the production of dust on rubbing, and by the aid of the magnifying glass it can be easily ascertained whether the covering or dressing is merely superficial or penetrates into the substance of the fabric.

The tests of permanence of dyes on fabrics are as follows:

Washing Fastness. Fabrics should stand mechanical friction as well as the action of soap liquor and the temperature of the washing operation. In order to test the fabric for fastness a piece should be placed in a soap solution similar to that used in the ordinary household, and heated to 131 degrees F. The treatment should be repeated several times. If the color fails to run it is fast to washing.

Fastness Under Friction. Stockings, hosiery yarns, corset stuffs, and all fabrics intended to be worn next [Pg 243] to the skin must be permanent under friction, and must not rub off, stain, or run, that is, the dyed materials must not give off their color when worn next to the human epidermis (skin), or in close contact with colored articles of clothing, as in the case of underwear.

The simplest test is to rub the fabric or yarn on white unstarched cotton fabric. In comparing the fastness of two fabrics it is necessary to have the rubbing equal in all cases.

Resistance to Perspiration. With fabrics coming in contact with the human skin it is necessary in addition to fastness under friction that they should withstand the excretions of the body. The acids of perspiration (acetic, formic, and butyric) often become so concentrated that they act on the fiber of the fabric.

In order to test the fabric for resistance, place the sample in a bath of 30 per cent dilute acetic acid (one teaspoonful to a quart of water) warmed to the temperature of the body, 98.6 degrees F. The sample should be dipped a number of times, and then dried without rinsing between parchment paper.

Fastness against Rain. Silk and woolen materials for umbrella making, raincoats, etc., are expected to be rainproof. These fabrics are tested by plaiting with undyed yarns and left to stand all night in cold water.

Resistance to Street Mud and Dust. Ladies' dress goods are expected to withstand the action of mud and dust. In order to test a fabric for this resistance the sample should be moistened with lime and water (10 [Pg 244] per cent solution), dried, and brushed. Or sprinkle with a 10 per cent solution of soda, drying, brushing, and noting any changes in color.

Fastness to Weather, Light, and Air. Various people have attempted to set up standard degrees of fastness—for every shade of color is affected by the action of sun, light, and air—and as a result fabrics that remain without appreciable alteration for a month of exposure to direct summer sunlight are classified as "fast," and those undergoing slight appreciable change under the same conditions as "fairly fast." "Moderately fast" colors are those altering considerably in fourteen days; and those more or less completely faded in the same time (fourteen days) are designated as "fleeting."

Directions for testing fastness of Color in Sunlight. Cover one end of the sample of cloth with a piece of cardboard. Expose the fabric to the sunlight for a number of days and examine the cloth each day in the dark and notice whether the part exposed has changed in color when compared with the part covered. Count the number of days it has taken the sunlight to change the color.

Brown in woolen materials is likely to fade. Brown holds its color in all gingham materials.

Dark blue is an excellent color for woolens and ginghams. Light blues on the other hand usually change.

Black, gray, and black with white. These colors are very satisfactory for woolen materials.

Black is not a color which wears very well with cotton fabrics, as it shows the starch (sizing) and often fades.

Red is an excellent color for all woolen materials. It looks attractive and wears well.

[Pg 245] Red is a very poor color for cotton. It loses its brilliancy and frequent washing spoils it.

A deep pink is an excellent color for all ginghams for it fades evenly and leaves a pretty shade.

Green is a poor color for both cotton and woolen materials unless it is high priced.

Lavender fades more than any other color in textiles.

History of Textiles

The three fundamental industries that have developed from necessity are the feeding, sheltering, and clothing of the human race. These primary wants were first gratified before such conveniences as transportation and various lines of manufacture were even considered. Next to furnishing our food supply, the industry of supplying clothing is the oldest and the most widely diffused. It is in the manufacture of textiles—including all materials used in the manufacturing of clothing—that human ingenuity is best illustrated.

The magnitude of the textile industry in the United States is evident when we consider that it gives employment to a round million of people, paying them nearly five hundred million dollars annually in wages and salaries, producing nearly one and three-quarters billion dollars in gross value each year, and giving a livelihood to at least three millions of our population.

Wool, cotton, flax, and silk have been used since early times. Even in the earlier days these fibers were woven with great skill. It is not known which fiber was the first to be used in weaving. It is probable, however, that the possession of flocks and herds led to the spinning [Pg 246] and weaving of wool before cotton, flax, or silk fibers were thus used.

Wool. The date at which prehistoric man discarded the pelt of skins for the woven fabric of wool marks the origin of the textile industry. Primitive sheep were covered with hair and the wool which now characterizes them was then a downy under-coat. As time went on and the art of spinning and weaving developed, the food value of sheep decreased, while the wool value increased. The hairy flocks were bred out, and the sheep with true wool, like the merino, survived. Sheep were bred principally for the wool and not for the mutton. Woolen fabrics were worn by the early inhabitants of Persia and Palestine. The Persians were noted for the excellent fabrics they wove from wool. Even the Hebrews of an early date were very skilful in weaving woolens.

The early Romans were a race of shepherds and the women of the higher classes wove the cloth in their own homes. When Caesar invaded England, he found in the southern part of the island people acquainted with the spinning and weaving of wool and linen. With the downfall of Rome, the art of weaving cloth in Europe was almost lost, and people again wore furs and skins.

By the end of the eleventh century English cloth manufacturing had begun to revive. In the northern part of Italy certain Italians had flocks of sheep and obtained very fine wool, and the people of Flanders continued to develop skill in weaving during the Dark Ages.

In the twelfth century the woolen manufacturers of [Pg 247] Flanders had grown to be of great importance, and some of the finest goods were shipped from there to many countries.

In England, up to the time of Edward III, in the fourteenth century, the wool produced was exported to the Netherlands, there to be woven into cloth. Edward III invited many of the Flemish weavers to come to England to teach the English people how to make their own clothes. Edward was called the "Royal Wool Merchant" and also the "Father of English Commerce." During Elizabeth's reign in the sixteenth century the chief article of export was woolen cloth. In 1685 the Huguenots, who were driven from France, went to England to settle. These people were noted for their skill in weaving.

Patient effort in care and breeding of sheep showed a steady increase in the quantity and quality of wool until 1810, and the proportion of sheep to the population was then greater than at the present time.

Our own climate is highly favorable for sheep breeding, and it is certain that the American sheep has no superior in any wool growing country, in constitutional vigor and strength of wool-fiber, and no wools make more durable or more valuable clothing.

The obstacles to sheep husbandry in certain parts of the United States, like New England, are mainly climatic. The natural home of the only races of sheep which can be herded in large flocks is an elevated tableland, like the steppes of Russia and the great plains of Asia, Argentina, Montana, Wyoming, and others of our [Pg 248] western states where an open air range is possible for nearly twelve months in the year. In these elevated lands there are grasses which are more nutritious in winter than in summer. The climate of New England does not permit the growth of such grasses. Every grass which will grow in New England becomes in the cold months frozen wood fiber. Then again there is the frigid and penetrating atmosphere which necessitates housing the sheep in winter, and these animals cannot be closely housed without engendering a variety of parasitic diseases.

Cotton. Long before history was written, cotton was used in making fabrics in India and China. Cotton has been for thousands of years the leading fabric of the East. The Hindoos have for centuries maintained almost unapproachable perfection in their cotton fabrics. It was the Arabian caravans that brought Indian calicoes and muslins into Europe.

Cotton was first cultivated in Europe by the Moors in Spain in the ninth century. In 1430 it was imported into England in large quantities. The section of England about Manchester became in time the seat of the great cotton industry; this was due to the settlement of spinners and weavers from Flanders.

During the reign of Elizabeth, the East Indies Trading Company was established. Not only was cotton imported, but also India muslins. This caused trouble because of the decrease in the demand for woolen goods manufactured in England. A law was passed prohibiting the importing of cotton goods and later the [Pg 249] manufacturing of them, but this law was repealed on account of the great demand for cotton materials.

Columbus found cotton garments worn by the natives of the West Indies. Later Cortez found that cotton was used in Mexico; hence, cotton is indigenous to America. In 1519 Cortez made the first recorded export of cotton from America to Europe.

In 1734 cotton was planted in Georgia. Bales of cotton were sent to England, and the manufacturing of cloth was soon under way. While the colonies were trying to gain independence, England imposed a fine on anyone sending cotton machinery to America, and restrictions were put on manufacturing and imports of any kind. After the War of Independence many of the southern states began to raise cotton in larger quantities.

The invention of the cotton-gin by Eli Whitney was one of the great inventions of the age. While only two pounds of cotton could be seeded by hand by one person in a day, the gin made it possible to do several hundred pounds. At the time of the Civil War the greater part of the cotton used by English manufacturers was imported from the southern states. The closing of the southern ports during the war affected the cotton industry throughout the world. Large mills in England were closed, and thousands of people were out of employment. Steps were then taken to encourage people of India, Egypt, Central and South America to increase their production of cotton, and from that time on, cotton from these countries has been found in the [Pg 250] general market. Cotton is now cultivated in nearly all countries within the limits 45° north and 35° south of the equator.

At the present time the United States ranks first in the production and export of cotton. Of all the states, Texas and Georgia produce the largest amount. About one-third of the entire crop is used in our own mills; $250,000,000 worth of cotton is annually exported, principally through New Orleans, New York, Savannah, and Galveston. Three-fifths of this quantity goes to mills in England; Germany, France, and Switzerland take a large part of the remainder.

The value of cotton is shown by the fact that about one-half the people of the earth wear clothing made entirely of cotton, and the other half (with the exception of some savage tribes) use it in part of the dress.

Linen. Linen has always been held in great esteem. The garments of the Egyptian, Hebrew, Greek, and Roman priests were made of the finest linen.

During the Middle Ages, Italy, Spain, and France were celebrated for their linen fabrics. Religious intolerance in France drove 300,000 of her best textile workers into England, Ireland, and Scotland. Irish linen weaving began as early as the eleventh century.

Linen has never been largely woven in America except in the coarser forms of crash and toweling, although linen weaving was one of the Puritan domestic industries. The reason America has not been able to equal Europe in its production of fine linens is because the process for separating the fiber from the stalk requires the cheapest [Pg 251] form of labor to make it profitable, hence most of the American-grown flax is raised only for seed.

Silk. Silk was used in the East as a fabric for the nobility. It was first used in China and later in India. It was brought into Europe about the sixth century. Up to that time the Chinese had a monopoly of the industry. By the tenth and eleventh centuries silk fabrics were made in Spain and Italy. At the close of the sixteenth century silk was being produced at Lyons, France. It was afterwards introduced into England, and the English silk for a long time replaced the French in the European market.

History of the Organization of Textile Industries

The development of the textile industry may be divided into four stages or periods: first, the family system; second, the guild system; third, the domestic system; and fourth, the factory system.

The Family System. Under the family system the work of spinning and weaving was carried on by members of a household for the purpose of supplying the family with clothing. There were no sales of the product. Each class in society, from the peasant class to that of the nobleman, had its own devices for making clothing. This was the system that existed up to about the tenth century.

The Guild System. As communities became larger and cities sprang up, the textile industry became more than a family concern. There was a demand for better [Pg 252] fabrics, and to meet this demand it became necessary to have a large supply of different parts of looms. The small weaver who owned and constructed his own loom was not able to have all these parts, so he began to work for a more prosperous weaver. The same conditions applied to spinning, and as early as 1740 spinning was carried on by a class distinct from the weavers. As a result the small weaver was driven out by the growth of organized capital, and a more perfect organization, called the guild system, arose. By this system the textile industry was carried on by a small group of men called masters, employing two, three or more men (distinguished later as journeymen and apprentices). The masters organized associations called guilds and dominated all the conditions of the manufacture to a far greater extent than is possible under present conditions.

It was the family system that existed in the American colonies at the beginning of the settlement, and for many years after. The guild system was not adopted in America because it was going out of existence on the Continent.

The Domestic Period. By the middle of the eighteenth century the textile industry began to break away from the guilds and spread from cities to the rural districts. The work was still carried on in the master's house, although he had lost the economic independence that he had under the old guild system where he acted both as merchant and manufacturer. He now received his raw material from the merchant and disposed of [Pg 253] the finished goods to a middleman, who looked after the demands of the market.

The Factory System. The domestic period was in turn crowded out of existence by the factory system. A factory is a place where goods are produced by power for commercial use. The factory system first came into prominence after the invention of the steam engine. No record has been found showing its existence prior to this invention.

English weavers and spinners became very skilful and invented different mechanical aids for the production of yarn and cloth. These mechanical aids not only enabled one man to do twenty men's work, but further utilization was made of water and steam power in place of manual labor. Then began the organization of the industry on a truly gigantic scale, combining capital and machinery and resulting in what is known as the factory system.

Previous to the development of the factory system there was no reason why any industry should be centered in one particular district. Upon the utilization of steam power the textile industry became subdivided into a number of industries, each one becoming to a great extent localized in convenient and suitable portions of the country. Thus in Bradford the wool of Yorkshire (England) meets the coal of Yorkshire and makes Bradford the great woolen and worsted center of the world. The same thing took place in Manchester, where the cotton of America meets the coal of England under satisfactory climatic conditions, and around [Pg 254] Manchester is the greatest cotton manufacturing of the world.

The same is true in America. Lawrence became a large worsted center on account of the great fall of water and the use of the river to deposit wool washings. Lowell, Fall River, and New Bedford became large cotton centers for similar reasons.

History of Manufacturing

SPINNING WHORL
One of the earliest devices used for spinning

Spinning. Spinning and weaving are two of the earliest arts practised by man. Yarn for the making of cloth was spun in the earliest times by the use of the distaff and spindle. The spindle was a round stick of wood a foot or less in length, tapering at each end. A ring of stone or clay was placed on the spindle to give it steadiness and momentum when it revolved. At the top of the spindle was a slit or notch in which the yarn was caught. The distaff was a larger, stouter stick, around one end of which the material to be spun was wound in a loose ball. The spinner fixed the end of the distaff under her left arm so that the coil of material was in a convenient position for drawing out to form the yarn. The end of the yarn, after being prepared, was inserted in the notch, and the spindle was set in motion by rolling it with the right hand against the leg. Then the spinner drew from the distaff an additional amount [Pg 255] of fiber, which was formed by the right hand into uniform strands. After the yarn was twisted, It was released from the notch and wound around the lower part of the spindle.

In order to spin yarn by the primitive spinner, it was necessary for the fiber to have sufficient length to enable it to be manipulated, drawn over, and twisted by the fingers. It is noted that the yarns for the gossamer-like Dacca muslins of India were so fine that one pound of cotton was spun into a thread 253 miles long. This was accomplished with the aid of a bamboo spindle not much bigger than a darning needle, which was lightly weighted with a pellet of clay. Since such a slender thread could not support even the weight of so slight a spindle, the apparatus was rotated upon a piece of hollow shell. It thus appears that the primitive spinners with distaff and spindle had nothing to learn in point of fineness from even the most advanced methods of spinning by machinery.

HAND SPINNING
From a Fourteenth Century MS. in the British Museum

Certain rude forms of the spinning wheel seem to have been known from time immemorial. The use of the wheel in Europe cannot, however, be dated back earlier than the fifteenth century. In the primitive wheel the spindle, having a groove worked in its whorl, was mounted horizontally in a framework fixed to the [Pg 256] end of a bench. A band passed around the whorl and was carried around a large wheel fixed farther back on the bench, and this wheel, being turned by the hand of the spinner, gave a rapid rotation to the spindle.

AN ANCIENT LOOM
From an Egyptian Monument

The fibers to be spun were first combed out by means of carding boards—an implement of unknown antiquity, consisting of two boards with wire teeth set in them at a uniform angle. The fiber to be carded was thinly spread upon one of the boards, and then the other was pushed backward and forward across it, the teeth of the two overlapping at opposite angles, until the fibers were combed out and laid straight in parallel lines. The fibers were then scraped off the boards in rollers or "cardings" about twelve inches long and three-quarters of an inch in diameter. An end of the carding was then attached to the spindle and the wheel set in motion. The carding itself was held in the hand of the spinner and gradually drawn out and twisted by the rotation of the spindle. As soon as a sufficient length had been attenuated and twisted to the required fineness, the thread so produced was held at right angles to the spindle and allowed to wind up on it. But for fine spinning two operations of the wheel were generally necessary. By the first spinning the fibers were drawn [Pg 257] out and slightly attenuated into what was called a roving, and by the second spinning the roving itself

221

passed through a similar cycle of operations to bring it to the required degree of attenuation and twist.

Many improvements in the primitive wheel were introduced from time to time. In its later developments two spindles were employed, the spinner being thus enabled to manipulate two threads at once, one in each hand. This was the latest form of the spinning-wheel, and it survived until it was superseded in the eighteenth century by the great series of inventions which inaugurated the industrial revolution and led in the nineteenth century to the introduction of the factory system.

EARLIER SPINNING AND WEAVING
From a Fifteenth Century MS. in the British Museum

Weaving. When or where man first began to weave cloth is not known, nor is it known whether this art sprang from one common center or was invented by many who dwelt in different parts of the world. There is such a sameness in the early devices for spinning and weaving that among some men of science it is thought that the art must have come from a common center.

[Pg 258] Fabrics were made on the farms two or three hundred years ago in the following manner: the men of the household raised the flocks, while the women spun the yarn and wove the fabrics. In this way the industry prospered, giving occupation and income to thousands of the agricultural class. You might say that in England fabrics were a by-product of agriculture. As time went on, farmers of certain sections of England became more expert in the art, and the weaving became separated from the spinning. The weavers became clustered in certain towns on account of the higher skill required for the finer fabrics. The rough work of farming made the hands of the weaver less skilful. This, coupled with the fact that the looms became more complicated with improvements, called for a more experienced man. Great inventions brought about a more rapid development of the factory.

Richard Arkwright, who has been called the "father of the factory system," built the first cotton mill in the world in Nottingham in 1769. The wheels were turned by horses. In 1771 Arkwright erected at Crawford a new mill which was turned by water power and supplied with machinery to accomplish the whole operation of cotton spinning in one mill, the first machine receiving the cotton as it came from the bale and the last winding the cotton yarn upon the bobbins. Children were employed in this mill, as they were found to be more dexterous in tying the broken ends. As the result of this great invention, factories sprang up everywhere in England, changing the country scene into a collection [Pg 259] of factories, with tall chimneys, brick buildings, and streets.

From 1730 to the middle of the nineteenth century the development of inventions was rapid:

1730 — First cotton yarn spun in England by machinery by Wyatt.

1733 — English patent granted John Kay for the invention of the fly shuttle.

1738 — Patent granted Lewis Paul for the spinning machinery supposed to have been invented by Wyatt.

1742 — First mill for spinning cotton built at Birmingham; moved by asses; but not successful.

1748 — Patent on a cylinder card as first used by hand, granted Lewis Paul.

1750 — Fly shuttle in general use in England.

1756 — Cotton velvets and quiltings first made in England.

1760 — Stock cards first used for cotton by J. Hargreave. Drop box invented by Kay.

1762-67 — Spinning-jenny invented by Hargreave.

1769 — Arkwright obtains his first patent on spinning.

1774 — Bill passed in England to prevent the export of cotton machinery.

1775 — Second patent of Arkwright on carding, drawing, and spinning.

1779 — Mule spinning invented by Crompton. Peele's patent on carding, roving, and spinning.

1782 — Date of Watt's patent for the steam-engine.

1783 — Bounty granted in England for the export of certain cotton goods.

1785 — Power loom invented by Cartwright. Cylinder printing invented by Bell. A warp stop-motion described in Cartwright's patent.

1788 — First cotton factory built in the United States, at Beverly.

[Pg 260] 1789 — Sea Island cotton first planted in the United States. Samuel Slater starts cotton machinery in New York.

1790 — First cotton factory built in Rhode Island by Slater.

1792 — First American loom patent granted to Kirk and Leslie.

1794 — Cotton-gin patented by Eli Whitney.

1801—Date given for invention of the Jacquard machine in France.

1803—Dressing machine and warper invented in England by Radcliffe, Ross, and Johnson.

1804—First cotton mill built in New Hampshire, at New Ipswich.

1805—Power loom successfully introduced in England after many failures.

1806—First cotton mill built in Connecticut, at Pomfret.

1809—First cotton mill built in Maine, at Brunswick.

1812—First cotton mill built at Fall River.

1814—Cotton opener with lap attachment invented in England by Creighton.

1815—Power loom introduced into the United States at Waltham.

1816—First loom temple of Ira Draper patented in the United States.

1818—Machinery for preparing sewing cotton invented in England by Holt.

1822—First cotton factory erected at Lowell.

1823—Differential motion for roving frames patented by Arnold. First export of raw cotton from Egypt to England.

1824—Tube frame or speeder patented by Danforth.

1825—Self-acting mule patented in England by Roberts.

1828—Ring spinning patented by John Thorpe. Cap spinning patented by Danforth.

1829—Revolving loom temple improvements patented by Ira Draper.

1832—Stop-motion for drawing frames invented by Bachelder.

[Pg 261] 1833—Ring spinning frames first built by William Mason.

1834—Weft fork patented in England by Ramsbottom and Hope. Shuttle-changing loom by Reid and Johnson.

1840 — Automatic loom led off. Important temple improvement.

1849 — First cotton mill erected in Lawrence.

Through this great change from hand to power work, thousands were thrown out of employment in the great textile centers, and much suffering occurred, which led to the smashing of machinery.

Knitting Machinery. Like many other industries, the hosiery trade owes its first and most important impetus to the genius of one who was not connected with the business in a practical way. This event took place when the Rev. William Lee invented the hand frame. He was married early in life, and his wife was obliged, on account of the slender family finances, to knit continuously at home. Struck with the monotony and toil involved in knitting with the hand pins, Mr. Lee evolved a means of knitting by machinery and brought out the hand stocking-frame, which to-day preserves its chief features very much as Lee invented them. When knitting by hand, one must form each loop separately, and loop follows loop laboriously until the width of fabric has been worked. Lee contrived to make the whole row of loops across the width simultaneously by arranging a needle for each loop and placing in connection with each needle a sinker and other apparatus for completing the formation of the loop. First of all, the yarn is laid over the needles, which are arranged [Pg 262] horizontally, and the sinkers come down on the yarn and cause it to form partial loops between the needles. The old loops of the previous course are now brought forward and the new yarn is drawn through them in the same way as is done on the hand pins. Thus the new yarn of one course is drawn through the loops of the preceding one, and so the whole fabric is built up. This frame of Lee's held its own in the great centers until some thirty years ago.

Lee's hand frame gave way to what is termed the jack and sinker rotary frame, which was like the hand frame in its chief features, but with the advantage that all the motions were brought about by power. The various operations were put under the control of a set of cams [20] and made to perform their movements in exactly the same way as in the case of the hand frame. In the first power machine for knitting, the machine builder used the cam mechanism, and in examining the latest machines we find that he has persisted in this course throughout. The cam movement is characterized by great smoothness of working and absence of vibration, which is very necessary in a machine of the delicate adjustment of the knitting frame. It is usual to connect some of the parts with two of these cams, one of which controls the up-and-down motion and the other the out-and-in movement. When these two cams work in conjunction, we obtain all the possible degrees of harmonic motion.

[Pg 263] From the jack and sinker frame the next really important step was taken when William Cotton brought out his famous Cotton's patent frame. In his machine the frame was in a sense turned on its back, for the parts, such as the needles, which had been horizontal, were made vertical and *vice versa*. He also reduced the number of the moving parts and perfected the cam arrangement. Another very important development of the machine was when it was built in a number of divisions so as to work a number of articles side by side at one time. At present there are knitting frames which can make twelve full-sized garments at one and the same time.

Another important improvement was effected when the fashioning apparatus was supplied to the machine, by means of which the garments could be shaped according to the human form by increasing or decreasing the width as desired.

History of Lace

Lace, like porcelain, stained glass, and other artistic things, has always been an object of interest to all classes. Special patterns of laces date from the sixteenth century. The church and court have always encouraged its production. While the early lace work was similar to weaving, in that the patterns were stiff and geometrical, sometimes the patterns were cut out of linen, but with the development of the renaissance of art, free flowing patterns and figures were introduced and worked in.

[Pg 264] The lace industry first took root in Flanders and Venice, where it became an important branch of industry. Active intercourse was maintained between the two countries, so that intense rivalry existed. France and England were not behind Venice and Flanders in making lace. The king of France, Henry III, encouraged lace work by appointing a Venetian to be pattern maker for varieties of linen needlework and lace for his court. Later, official aid and patronage were given to this art by Louis V. Through the influence of these two men the demand for lace was increased to such an extent that it became very popular.

Under the impulse of fashion and luxury, lace has received the stamp of the special style of each country. Italy furnishes its Point of Venice; Belgium its Brussels and Mechlin; France its Valenciennes, etc.

Very little is known of the early lace manufacturers of Holland. The laces of Holland were overshadowed by the richer products of their Flemish neighbors. The Dutch, however, had one advantage over other nations in their Haarlem thread, once considered the best thread in the world for lace.

In Switzerland, the center of the lace trade, the work was carried on to such a degree of perfection as to rival the laces of Flanders, not alone in beauty, but also in quality.

Attempts have been made at various times, both during this century and the last, to assist the peasantry of Ireland by instruction in lace-making. The finest patterns of old lace were procured, and the Irish girls [Pg 265] showed great skill in copying them. Later a better style of work, needlepoint, was modeled after old Venetian lace — the exquisite productions for which Americans pay fabulous prices at the present day.

The lace manufacturers of Europe experienced a serious set-back in 1818 when bobbinet was first made in France. Fashion, always fleeting, adopted the new material. Manufacturers were forced to lower prices, but happily a new channel for export was opened in the United States.

The machine-made productions of the Nottingham looms, as triumphs of mechanical ingenuity, deserve great praise.

The first idea of the lace-making machine is attributed to a common factory hand, Hammond Lindy, who, when examining the lace on his wife's cap, conceived a plan by which he could copy it on his loom. Improvements followed, and in 1810 a fairly good net was produced.

Perhaps the most delicate textile machine known, in its sensitiveness to heat and cold, is a lace machine. A machine can be made to run in any climate, provided it is so installed as to be protected from either extreme of temperature.

The various substitutes for hand-made lace are legion; for what the inventor cannot achieve in one way he can in another. There remains however the fact that the productions of machinery can never possess the charm of the real hand-made work. Machine-made lace is stiffer than hand-made lace.

FOOTNOTES:

[19] The testing apparatus may be obtained from any textile manufacturing company, such as Alfred Suter, 487 Broadway, New York.

[20] A cam is a device consisting of a special shaped wheel attached to a machine to give a special kind of motion or movement.

[Pg 267]

EXPERIMENTS

Experiment 1 — Construction of Cloth

Apparatus: Pick glass, dissecting pin, [21] foot-rule.
Materials: 4 square inches of burlap.
References: *Textiles.* See page **54** , Weaving; page **1** , Fibers.

Directions

1. Look at the cloth under the pick glass and describe the appearance and structure of its interlacing threads, called weave.

2. With a pin separate the interlacing threads of the cloth which are called *warp* and *filling*. *Warp* is composed of yarn running in the direction of the length of the cloth. *Filling* is composed of yarn running at right angles to the warp.

a. What are the interlacing threads of cloth called?

b. Of what is warp composed and in what direction do the warp threads extend? filling?

3. Notice the appearance of the individual threads (called *yarn*) of the warp and filling. Test the strength of the yarn by trying to break it.

4. Untwist one of the warp threads and one of the filling threads. Notice whether the yarn becomes stronger or weaker as it is untwisted. What effect has twist on the yarn?

5. After untwisting one of the threads what remains? Measure the length of several of these ends called *fibers*. Describe the appearance of the fiber as to curl, feel, fineness, etc.

Questions

1. Of what does yarn consist?

2. What causes the fibers to cling together?

3. What is the process called by which two sets of threads interlace?

[Pg 268] 4. When two sets of threads interlace or are woven what is produced?

Experiment 2 — Plain or Homespun Weave

Apparatus: Hand loom, [22] two pencils, scissors.
Material: Yarn of two colors.
Reference: *Textiles*, page **58** .

Directions

1. Make a warp on the hand loom with green yarn by having parallel threads running the longest way of the loom to the notches.

2. A *harness* is a framework on a loom used for raising certain warp threads. Use a pencil as a harness and raise the 1st, 3d, and 5th warp threads. A *shed* will in this way be formed through which the shuttle carrying the filling thread will pass. Use the red yarn for filling and attach it at one end before passing it through the shed.

3. With a second pencil to act as a second harness raise the 2d, 4th, and 6th warp threads. Pass the filling through the shed thus formed.

4. Repeat twice Directions 2 and 3.

5. Tie all ends, cut the woven sample away from the loom, and mount in note-book.

Questions

1. What part of a loom is the harness?

2. What is meant by a shed?

3. What carries the filling thread through the shed on a loom?

4. What is the principle of plain weaving?

5. Name some fabrics produced by plain weaving? See *Textiles*, page **58** .

Experiment 3—Twill Weave

Apparatus: Hand loom, four pencils, scissors.
Materials: White cotton warp, colored yarn filling.
Reference: *Textiles*, page **58** .

[Pg 269]

Directions

1. On the hand loom make a warp by threading four white warp threads to a notch until there are six sets of warp threads.

2. Using a pencil as a harness (See Exp. **2**) raise the first thread of each set of warp threads and pass the filling thread through the shed thus formed.

3. With another pencil as a second harness raise the second thread of each set of warp threads and pass the filling.

4. With a third pencil raise the third thread of each set of warp threads and pass the filling.

5. With still another pencil to act as a fourth harness raise the fourth thread of each set and again pass the filling.

6. Repeat the above directions (2 to 5) several times. Notice that the moving of the filling thread, one warp thread to the left, each time it is woven is causing a diagonal line or rib to form, called *twill*.

7. Cut the woven sample away from the loom and mount.

Questions

1. Why is this weave called a twill weave?

2. How is the diagonal line or twill formed?

3. Why would this kind of weaving be spoken of as 4-harness weave?

4. What popular dress fabric is of twill weave?

Experiment 4 — Comparison of Plain and Twill Weave

Apparatus: Pick glass, dissecting pin, foot-rule.
Material: 4 sq. in. of burlap, 4 sq. in. of serge.
References: *Textiles*, pages **58** , **59** , **60** .

Directions

1. Examine the burlap under the pick glass, noting the structure and number of threads to the inch in the warp (called *ends*) and the number of threads to the inch in the filling (called *picks*). Verify with foot-rule.

2. Repeat the above, using serge.

[Pg 270]

Questions

1. What is meant by a number of "ends to the inch"? a number of "picks to the inch"?

2. How many ends to the inch in the burlap? How many picks to the inch?

3. How many ends to the inch in the serge? How many picks?

4. Note several differences between cloth produced by plain weaving and cloth produced by twill weaving.

Experiment 5 — Pile Weave

Apparatus: Hand loom, two pencils, scissors.
Materials: White cotton warp, filling yarn of two colors.
Reference: *Textiles*, page **62** .

Directions

1. Thread the loom two warp threads to a notch until there are 20 ends (warp threads).

2. Use a pencil as a harness. Raise the 1st, 3d, 5th, 7th, and 9th sets of warp threads.

3. Fasten securely the green filling yarn at one end and pass it through the shed formed by carrying out Direction 2. Draw the filling thread tight and wind once or twice around the outside warp end.

4. Use a second pencil as a harness and raise the sets of warp threads that are now down, forming a new shed.

5. Fasten the red filling yarn at one end and pass it through the shed. Wind once or twice about the outside warp end.

6. Raise the red filling to form a loop in each place where it (the red filling) has passed over a warp end.

7. Form a shed by raising the first harness and pass through the green filling thread, drawing it tight to hold the red filling above it in place. Wind about the outside warp end.

8. Repeat Directions 2-7 several times, each time raising the red filling to form loops and each time drawing the green filling tight to hold the red in place.

9. Cut with scissors the loops formed by raising the red filling.

[Pg 271] 10. As well as you can with scissors, shear the pile (the soft, thick covering on the face) to make a fairly even surface.

11. Cut the sample away from the loom and mount.

Questions

1. What are some varieties of cloth that are woven with a pile surface?

2. Sometimes the loops of the pile are cut and sometimes left as loops. What fabrics are examples of cut pile? uncut pile?

3. What is meant by the *pile* of velvet or carpet?

Experiment 6 — Other Classes of Weave

Apparatus: Pick glass, dissecting needle.
Materials: Samples of satin, voile, lace curtaining, double cloth, carpeting.
Reference: *Textiles*, pages **58-64** .

Satin Weave

1. Examine the sample of satin under the pick glass. Notice that the warp and filling interlace in such a way that there is no trace of the diagonal on the face of the cloth.

a. Is satin of a close or loose weave?

b. What can you say of the surface of satin?

c. What effect has this smooth surface on light?

d. This is called a satin weave. Why?

e. What is the most extensive use of the satin weave? (See *Textiles,* page **1** .)

Note.—Sometimes fabrics of other weaves will have a satin stripe.

Gauze Weave

2. Examine the sample of voile under the pick glass. This is a type of what is known as gauze weave.

a. What is the chief characteristic of the gauze weave?

b. Name several gauze fabrics.

[Pg 272]

Lappet Weave

3. Examine a piece of lace curtaining under the pick glass.

a. If the fancy figures were not present, of what weave would this sample be?

Simple figures are stitched into plainly woven or gauze fabrics by machinery to imitate embroidery. This style of weave is known as lappet weave.

b. On fabrics of what two weaves is lappet weaving used?

c. What is lappet weaving?

Jacquard Weave

4. Examine a piece of carpet. Notice the elaborate designs or patterns and the number of colors used. When the figures are elaborate

236

they cannot be stitched in by simple lappet weaving. A special attachment called the *Jacquard* apparatus is placed on top of the loom. This apparatus controls the warp threads so that a great many sheds may be formed and elaborate figures woven into fabrics. This is called Jacquard weaving.

    a. What must be added to a loom for Jacquard weaving?

    b. What is the use of the Jacquard apparatus?

    c. When is the Jacquard weave used instead of lappet weave?

    5. Read *Textiles*, page **61** .

Double Cloth Weave

    6. Examine the sample of double cloth. Notice that there are two single cloths. They are combined into one by here and there lacing the warp and filling of one cloth into the warp and filling of the other. In this way they are fastened together securely.

    a. What color is the sample on one side? the other?

    b. Of what is double cloth composed?

    c. How are the single cloths combined into one?

    d. Read *Textiles*, page **62** . What are some of the uses of double cloth?

Classes of Weave

    7. How many classes of weave have been studied?

    8. Name the classes of weave.

    9. Name a fabric to illustrate each weave.

[Pg 273]

Experiment 7—Fibers

Apparatus: Pick glass, dissecting needle.
Materials: Samples of broadcloth, mohair, silk, cotton cloth, linen.
References: *Textiles*, pages **1** ; **97** , Mohair; **203** , Silk; **105** , Cotton; **193** , Linen; **199** , Hemp; **201** , Jute; **232** , Ramie; **233** , Pineapple.

1. Read *Textiles*, page **1** , paragraph 1. What are textiles?

2. Cloth is composed of yarn. Yarn in its turn is composed of many small ends called fibers.

3. Look at the sample of broadcloth. If you did not know this to be broadcloth you would speak of it as woolen goods. Detach from the sample a filling thread and separate it into fibers. These are woolen fibers.

4. Examine the sample of mohair and separate a filling thread into fibers. This takes the name mohair from the fibers which compose it. Mohair is obtained from the Angora goat.

5. Examine a sample of silk, also a detached filling thread. The silk fiber consists of a thread spun by the silk worm.

6. Wool, mohair, and silk fibers are obtained from the animals, the sheep, goat, and silk worm, hence they are called animal fibers.

7. Detach from the sample of cotton cloth a filling thread and separate it into fibers. These are cotton fibers and are obtained from the cotton plant.

8. Examine the sample of linen, a filling thread and its fibers. Linen is composed of fibers obtained from the flax plant.

9. Cotton and linen fibers are obtained from plants, and are called vegetable fibers. There are other vegetable fibers such as jute, hemp, ramie, pineapple, etc., but cotton and linen are the most important.

10. Name the most valuable fibers for textile use.

Questions

1. Of what is cloth composed?

2. Of what does yarn consist?

[Pg 274] 3. How are the fibers made to join in one long thread? (See Experiment **1** .)

4. Of what fibers are woolen and worsted goods composed?

5. Of what animal is wool the covering?

6. Of what fibers is mohair composed?

7. From what animal is mohair obtained?

8. Of what does the silk fiber consist?

9. What are the animal fibers?

10. Why are they called animal fibers?

11. Of what fibers is cotton cloth composed?

12. From what plant are cotton fibers obtained?

13. From what plant is the linen fiber obtained?

14. What are the most important vegetable fibers?

15. Name four other vegetable fibers.

16. Why are these fibers called vegetable fibers?

Experiment 8 — Wool Fiber

Apparatus: Pick glass, microscope, 2 pine cones, foot-rule.
Materials: Raw wool, woolen yarn.
Reference: *Textiles*, chapter **i** .

Directions

1. Separate a strand of woolen yarn into fibers. Examine both these fibers and fibers pulled from the raw wool. Would you describe these fibers as coarse or fine?

2. How do the fibers feel to touch?

3. Test the strength of the wool fibers by trying to break them.

4. Measure the length of several fibers.

5. Why was it difficult to straighten the fibers to measure them?

6. Extend the fiber to its full length, then release. How does this prove the fiber to be elastic?

7. Examine the fibers under the microscope. Describe. Notice that the wool fiber is cylindrical in shape. Notice that it is covered with

scales which overlap much as do the tiles of a roof or the spines of a pine cone.

8. Hold one pine cone with the spines pointing upward. With the spines of the other pointing downward press the second cone [Pg 275] down on the first. What happens? Just so the scales or points of the wool fibers hook into one another and interlock. These scales or serrations give to the wool fiber its chief characteristic which is the power of interlocking known as *felting* or *shrinking*.

9. See *Textiles*, page **2** , the drawing of a magnified wool fiber. Make a drawing of a wool fiber.

10. Examine under the microscope a hair from your head. Wool is only a variety of hair. Notice that the scales on the hair lie close to the stem and do not project as in the woolen fiber, hence hair fibers cannot interlock as wool fibers do. The scales lying close to the hair give a smooth surface to the fiber and make luster a characteristic.

11. Compare the wool fiber with hair, noting two differences.

Questions

1. With what is the wool fiber covered?

2. Of what advantage are these scales or points?

3. What is the chief characteristic of wool?

4. What is meant by the shrinking or felting power?

5. Name five characteristics of the wool fiber.

Experiment 9—Mohair Fiber

Apparatus: Microscope, foot-rule.
Materials: Wool fibers, mohair fibers, sample of mohair brilliantine.
References: *Textiles*, pages **1** , **37** , **97** .

Directions

1. Pull a mohair fiber from the fleece. Hold it up to the light. Describe the fiber as you see it.

2. Hold a mohair fiber and a wool fiber side by side to the light. Note the differences.

3. Measure several mohair fibers.

4. Examine the mohair fiber under the microscope. The fiber is covered with scales, but they lie close to the fiber and do not project in points as do the scales on the wool fiber, hence mohair will not felt to any degree.

[Pg 276] 5. The Angora goat of Asia Minor furnishes the mohair. This goat is being raised in the western states of the United States now.

6. Detach from the sample of mohair brilliantine a warp thread; a filling thread. Which is mohair? Which is cotton?

7. What word would describe the feel of mohair brilliantine? the appearance?

8. What are the characteristics of the mohair fiber?

9. What are the uses of mohair? Mohair is used in the manufacture of plushes, dress fabrics, and imitation furs.

Questions

1. Why will mohair not felt as wool does?

2. The scales lying close to the stem will have what effect on the surface of the fiber?

3. What effect will a smooth surface have on light?

4. What characteristic is given to mohair from the fact that the smooth surface reflects light?

5. From what animal is mohair obtained in greatest quantity?

6. Where is mohair being grown in the United States?

Experiment 10 — Cotton Fiber

Apparatus: Microscope, foot-rule.
Materials: Tuft of cotton fibers, cotton ball, seeds.
Reference: *Textiles*, chapter **ix** , page **105** .

Directions

1. Hold a tuft of cotton fibers tightly between the fingers and thumb of each hand and pull apart with a jerk. What is your judgment of the strength of the *staple* (fiber)?

2. Loosen gently the fibers of one of the tufts you have pulled apart. What is the feel of cotton? the appearance as you hold it to the light?

3. Detach several fibers one by one. How does the length compare with that of the wool and mohair? Measure and record the length of three fibers.

[Pg 277] 4. How do cotton fibers compare in fineness with wool fibers?

5. Compare the elasticity of cotton with that of wool.

6. Examine the cotton fibers under the microscope. Observe that the enlarged fiber looks like a twisted ribbon. When the fiber was growing it was cylindrical in shape. When ripe the plant drew back its life-giving fluid from the fiber and it collapsed and twisted like a corkscrew. The twist is peculiar to the cotton, being present in no other fiber. The twist makes the cotton fiber suitable for spinning, helping to hold the short fibers together.

7. Read of the cotton plant from *Textiles*, chapter **ix** .

8. The four chief cotton producing countries are the United States, Egypt, India, Brazil.

9. There are several classifications of cotton. The most common are Sea Island (in the lead); Egyptian (a close second); Uplands (that of the United States, southern part); and Peruvian.

10. Uplands is the most common cotton of our South.

Questions

1. What characteristic causes the cotton fiber to be easily recognized under the microscope?

2. Why does the twist render the cotton fiber suitable for spinning?

242

3. What are the characteristics of the cotton fiber?

4. Why is cotton known as a vegetable fiber?

5. Name the chief cotton producing countries.

6. What are the most common classifications of cotton?

7. What is the finest growth of cotton? (Sea Island commands at the present time $1.00 a lb., while Middling Uplands brings 15 cents.)

8. Where is cotton known as Upland Cotton grown?

Experiment 11 — Silk Fiber

Apparatus: Tripod, alcohol lamp, small pan of water, lead pencil.
Material: Silk cocoon.
Reference: *Textiles*, chapter **xvii** , page **203** .

[Pg 278]

Directions

1. Place the cocoon in a small pan of water. Apply heat to the pan until the water boils. The cocoon is placed in hot water to soften the glue which holds the fibers together.

2. Remove the outside loose fibers which cannot be reeled. This tangled silk on the outside of the cocoon is called *floss*.

3. Brush the finger over the cocoon to find the loose ends. Unwind carefully until you find a continuous end. Wind or *reel* the silk fiber over a lead pencil.

4. The silk fiber is the most beautiful and perfect of all fibers.

5. Hold the cocoon to the light as you reel. How does the silk fiber compare in fineness with the wool and cotton fibers?

6. The silk fiber is from 1000 to 4000 feet long. Unlike the other fibers the silk fiber is already a thread.

7. How does light effect the silk fiber? When the gum is thoroughly washed off the silk takes on its luster which is its chief characteristic.

8. Break the fiber after you have reeled a small quantity. Notice how the fiber springs back. Extend and release again. What characteristic does this illustrate?

9. Examine the silk fiber under the microscope. Notice that it is round and smooth and resembles a glass rod. It shows what appear to be two fibers united by the gum secreted at the same time that the fiber was formed. Describe the silk fiber as it appears under the microscope.

10. Silk is taken from the reel and twisted into a skein of raw silk and thus exported.

11. The manufacture in the United States begins with raw silk. It is handled here first by the *throwster* who winds it from the skein and makes different varieties of thread.

Questions

1. Why is the silk cocoon first placed in hot water?

2. What is known as floss?

3. What is meant by silk reeling?

4. What can you say of the length of the silk fiber?

5. In what way does the silk fiber differ from the other fibers?

[Pg 279] 6. What is the chief characteristic of the silk fiber?

7. What are other characteristics of the silk fiber?

8. In what form is silk exported?

9. In what countries is most of the raw silk produced? (See *Textiles*, page **206** .)

10. With what does the silk manufacture in the United States begin?

11. Who is the *throwster* and what is his work?

Experiment 12 — Linen Fiber

Apparatus: Microscope.
Material: Flax fibers.

Reference: *Textiles*, chapter **xv** , page **193** .

Directions

1. The linen fiber is obtained from the flax plant. Certain fibers, such as flax, jute, and ramie, are obtained from the stem of the plant, hence are known as *bast* fibers, and flax is the most important bast fiber.

2. It is difficult to separate the flax or linen fiber from the woody part of the stem. The process is called *retting*, which is really rotting by soaking the stem in water.

3. Before the fibers are entirely free from the woody part of the plant they undergo the processes of beating, breaking, scutching, hackling, etc.

4. Read the account of each process. See *Textiles*, pages **194** , **195** .

5. Measure and record the length of two linen fibers.

6. Test the strength by trying to break the fiber.

7. Test for elasticity.

8. What is the appearance of the linen fiber when held to the light?

9. What is the color of the fiber? What is the process called by which linen is whitened? (Bleaching.)

10. Examine the flax fibers under the microscope. Observe that the fibers look like long cylindrical tubes. Describe the appearance of linen fibers under the microscope.

11. The best flax is grown in Belgium and Ireland.

[Pg 280]

Questions

1. From what part of the plant are bast fibers obtained?

2. Name some bast fibers.

3. What is the most important bast fiber?

4. What is retting?

5. For what purpose is linen subjected to retting?

6. Through what five processes does the flax fiber pass before it is free?

7. Where is the best flax grown?

Experiment 13 — Carding

Apparatus: A pair of hand cards.
Material: Small quantity of scoured wool.
References: *Textiles*, pages **39** and **50** .

Directions

1. Examine the hand cards. Notice that there is a foundation of several layers of leather. Notice that this foundation is covered with staples of steel wire. Notice that the staples are shaped like the letter U with the points turned one way. The covering of the hand cards is called *card clothing*.

2. Hold one hand card in the left hand, face up, wires pointing to the left. Spread the wool over the pointed wires of this card.

3. Hold the other card in the right hand, face down, with the wires pointing to the right. Bring the pointed wires of this card down on the wool and drag it lightly through the wires of the other card. Repeat several times.

4. You have been *carding* wool. The sharp points have been tearing the wool apart or disentangling the fibers. Carding brushes the fibers out smooth and makes them somewhat parallel. It forms them into a thin sheet.

5. The wool must be carded many times before it is sufficiently disentangled for drawing and spinning. In order to card again the hand card must be *stripped* of the wool so that it may be dragged again through the staples.

6. Hold the hand card, which is in your right hand, erect. Notice that the wires point downward. Move the other hand [Pg 281] downward over the wires. Notice that the surface is smooth. The

points do not prick as they will if you try to brush the hand up-wards over the wires.

7. Hold the card in the left hand in a similar position. Raise and bring the sharp wires of this card down on the smooth surface of the other card and strip it of its wool.

8. Card again, then strip again. Repeat several times until the fibers are thoroughly disentangled.

9. This carding and stripping, once done by hand, is now done in the mill by a power machine called the *card*. (See picture, *Textiles*, page **38** .) Notice that instead of cards this machine consists of rollers or cylinders. Some are carding cylinders and some are stripping cylinders. The principle is the same as that of the hand cards. The wool is carded and stripped again and again and is finally delivered in a soft, fluffy rope called a *sliver* ready for drawing and spinning.

Questions

1. What is the covering of the hand card called?

2. Describe card clothing.

3. What does carding do to the wool?

4. When the sharp wires of one cylinder meet the sharp pointed wires of another cylinder what is the action on the wool?

5. If the sharp points of one cylinder meet the smooth surface of another cylinder what happens to the wool on that cylinder?

6. In what form does the wool finally leave the machine? What name is given to this fluffy rope?

7. How was carding done in the early days? How is it done now?

8. In what way is the principle of the hand cards the same as that of the card of the mill?

Experiment 14 — Drawing and Spinning

Apparatus: Foot-rule, elastic band.
Material: Small quantity of scoured wool.
References: *Textiles*, pages **4** , **44** , **134** ; Sections: Spinning: Mule

Spinning.

Directions

1. Observe the mass of wool fibers. The wool was clipped from the sheep, *washed*, and *oiled* to make it smooth and pliable.

2. With the fingers gently open up or loosen the mass of wool fibers. In the mill this is done by a machine called the *card*. (See picture, *Textiles*, page **38** .) And the process itself is called *carding*.

3. Gently *draw* out the mass of fibers until you have drawn it into one long strand.

4. Draw it again and again until to draw it would cause it to break.

5. This process in the mill is known as *drawing*. The wool passes through machine after machine, which gradually reduces the thickness of the strand.

6. You have now a strand called *roving*, but not a thread with which you could weave. What is called the strand? Why could you not weave with it as it is? If you pulled the roving apart it would separate into a number of small ends. What name is given to these ends?

7. It is necessary to hold these fibers together in a thread. Hold the roving in the left hand and with the right hand draw the fibers out several inches. As you draw, twist the roving between the fingers and thumb. The *twisting* is called *spinning*.

8. When you have twisted sufficient yarn to attach to the end of a foot-rule, do so. Give a whirl to the ruler, which is taking the place of the old-time *spindle*, and let it drop. Continue to whirl the ruler and notice that as it revolves the yarn is twisting. When well twisted, wind the yarn on the ruler. There was a hook on the old-time spindle. Instead of the hook, hold the wound yarn in place by an elastic band. Draw out several inches again and repeat.

9. With the spindle a *distaff* was used. It held the roving which you now hold in your left hand. (See **picture** of distaff and spindle.)

248

10. Define spinning; see *Textiles*, page **4** , footnote. The early use of the spindle was the same as its use of to-day. In what two ways is the spindle of use?

11. The improvement on the distaff and spindle was the spinning wheel. Now the spinning frame in the mill has replaced both.

[Pg 283]

Questions

1. After shearing, through what two processes does wool pass?

2. Why is it necessary to oil wool?

3. What is the work of the *card*?

4. Explain the process called *drawing*. Why is it necessary to repeat the operation several times?

5. What followed the distaff and spindle in the development of spinning?

6. On what is the spinning done now in the mill? See *Textiles*, picture, pages **135** , **137** .

Experiment 15 — Gilling and Combing

Apparatus: Coarse comb, fine comb.
Material: Small quantity of scoured wool.
Reference: *Textiles*, pages **39-44** .

Directions

1. Open up the wool a little with the fingers. Do this in place of carding, as you need but a small quantity.

2. You comb your hair to make the hairs lie parallel, side by side, in place. Combs are used on wool for just the same purpose, but the first process of combing is not known as such. It is called *gilling*, and the combs themselves are called *fallers*. The machines are known as *gill boxes*. See *Textiles*, page **43** .

3. Hold the carded wool in the left hand in the middle of the strand. With the coarse comb in the right hand, comb and thus

straighten the fibers first at one end then at the other. This is *gilling.* The principle of gilling is to comb the fibers more and more nearly parallel and to draw them out into more even strands.

4. The coarse comb causes the hairs to lie parallel. A fine comb will further straighten the hairs, but it will also remove the snarled, tangled, short hairs. Again wool is to be treated like hair. Hold the strand in the middle as before. Comb each end with the fine comb. Notice that the fine comb is removing the short fibers and leaving the long fibers between the fingers. This is the second process of combing, and is called *combing.*

5. The long fibers are called *tops* and the short fibers are known [Pg 284] as *noils.* [23] Combing is the process which separates the long fibers known as *tops* from the short fibers known as *noils.*

6. The combing machine in the mill is a very complicated one. See picture, *Textiles*, page **41** .

7. Gill and comb several strands of wool.

8. Top is too delicate, as it comes from the comb, to be handled. The next process is to combine several strands into one. Combine the several strands you have gilled and combed. Comb this one end with the coarse comb again to be sure that the fibers are perfectly parallel.

9. You gilled, combed, and gilled again. So it is in the mill. After combing, the wool is gilled again by machines known as *finisher gill boxes,* and wound into a ball called *a top.*

10. *A top* differs from *top. Top* is the strand of long fibers which comes from the comb. *A top* is the ball of combed wool as it comes from the finisher gill boxes. It weighs from 7 to 12 lbs. and contains 200 to 250 yds.

11. The wool is now ready for the next processes—those of drawing and spinning.

Questions

1. Why is the hair combed? Why is wool combed?

2. What is the first process of combing called? What name is given to the combs used in gilling? What are the machines called?

3. What is the principle of gilling?

4. How does a fine comb act on the hair?

5. When you combed the wool with the fine comb what happened?

6. What are the long fibers called? the short? Of which are there more?

7. What is the second process of combing called?

8. Why is it necessary to combine several strands of top into one end?

9. Why is it necessary to gill again after combing?

10. In what form does the wool finally leave the finisher gill boxes?

11. What is a top?

[Pg 285] 12. What two processes follow carding?

13. For what two processes is wool now ready?

Experiment 16 — Raw Wool to Yarn

Apparatus: Hand cards, coarse and fine combs, pencil.
Material: Scoured wool.
Reference: The preceding experiments.

Directions

1. This wool has already been subjected to the three operations of shearing, scouring, and oiling.

2. Card the wool. What does carding do to the wool?

3. Strip the cards. Rub the sheet of fibers between the palms of the hands into the form of a strand. It is in this form that it leaves the card of the mill, and it is known as a *sliver* of wool.

4. Pull about three inches of wool from the sliver and perform upon it the operation of gilling by combing it with the coarse comb.

5. Follow the gilling by the operation of *combing*, which you will do by combing again, this time with a fine comb.

6. Pull about three inches again from the sliver. Continue to gill and comb by section until the entire sliver has been gilled and combed.

7. Combine several strands into one and subject the one strand to a second process of gilling to make sure that all fibers are side by side.

8. Gently draw out this strand of combed long fibers known as top. As you draw, spin. As you spin, wind on a lead pencil. The fineness of the yarn depends on the amount of drawing and twisting.

9. What is the source of wool? You began with wool, covering of the sheep's body, and after subjecting it to a series of operations you have converted it into yarn which is ready for weaving.

10. Name the operations in order, through which raw wool passes before it finally becomes yarn.

[Pg 286]

Questions

1. What are the first three processes through which wool passes? What is shearing? scouring?

2. Why is wool oiled?

3. What is meant by a sliver of wool?

4. What does gilling do to the wool?

5. What does combing do to the wool?

6. Why is there another operation of gilling after combing?

7. What is meant by *drawing*? *spinning*?

8. What name is given to the wool wound on the pencil?

9. On what does the fineness of the yarn depend?

Experiment 17 — Difference between Woolen and Worsted Yarn

Apparatus: Pick glass.
Materials: Sample of woolen cloth and worsted cloth.
References: *Textiles*, pages **50** and **51** .

Directions

Take a piece of worsted fabric and separate a piece of yarn from either the warp or filling. Do the same with a piece of woolen fabric. Notice the appearance of each piece of yarn. Which is smoother? What effect would friction have on the worsted yarn? the woolen yarn? Which sample of yarn would shine and reflect the light?

Experiment 18 — Burling and Mending

Apparatus: Chalk, scissors, dissecting pin, needle, pick glass.
Material: 4 square inches of cloth from the loom.
Reference: *Textiles*, page **71** .

Directions

1. Cloth from the loom is far from being a finished product. It must pass through several processes before it is finished. These processes are known as *finishing*.

2. What is the feel of this cloth?

[Pg 287] 3. Hold the cloth to the light and look through it. Note the imperfections and chalk them. What defects did you notice?

4. Place the cloth on the desk, face down. Rub the fingers over the back of the cloth. When the fingers locate a knot, raise it with the dissecting needle to be cut off later.

5. Reverse the cloth. Rub the fingers over the face. When a knot is found, force it through to the back with the dissecting needle. All the knots are on one side now. Clip them off with the scissors. This is called *burling* and is the first process of finishing.

6. Hold the cloth to the light. Notice where an entire filling thread is missing. This is known as a *full miss pick*. When part of a filling thread is missing it is spoken of as a *half miss pick*. In general what does a miss pick mean?

7. Unravel a filling thread from the lower edge of the cloth. With it thread a needle and replace the missing pick. Follow the weave closely, using a pick glass as an aid. You are performing the second process of finishing, that of mending.

8. If a warp end is missing replace it.

Questions

1. What is meant by *finishing*?

2. What is the first process of finishing? What is burling?

3. What is a full miss pick? a half miss pick?

4. What is the second process of finishing? What is mending? Of what must the mender be careful?

Experiment 19 — Removal of Stains

Material: Stained fabrics.

Textiles are easily stained, therefore it is necessary to know something about the character of stains and the methods of removal. Stains may be roughly divided into the following classes:

a. Stains from foods, such as grease and fruit acids.

b. Stains from machinery, as wheel grease and oils.

c. Blood stains.

d. Inks.

e. Chemicals, such as acids, alkalies.

Food stains are usually due either to grease contained in soup, [Pg 288] meat, milk, etc., or to sugar contained in candies or preserves, or to fruit acids contained in fresh fruits or sauces.

Wheel grease and lubricant stains are obtained from various parts of machines, like elevators, street cars, etc. After the cloth leaves the loom it often contains spots of grease, oil, or dirt stains due to drippings from the loom or overhead machinery. These are removed by means of liquids called solvents that dissolve the stain. Ether is the principal solvent used in the mill to remove small stains.

Very few people realize that vapors of cooked food and fat, unless carried out of a house, will condense and settle on fabrics in the form of a film which collects a great deal of dust. (A bad grease spot usually has a neglected grease spot for a foundation.) In order to break up this film it is necessary to separate the entangled dust. This is performed by some mechanical means, such as shaking and brushing.

The most effective method of removing a stain is to place a circle of absorbent material [24] around the spot to take up the excess of liquid. A white cloth should be placed under the fabric to absorb the solvent and show when the goods are clean. Then apply the solvent with a cloth of the same color and texture (satin is excellent as it does not grow linty) and rub from outside the spot to the center to prevent spreading. It is necessary to rub very carefully as excessive rubbing will remove the nap and change the color. One of the great dangers in removing a stain is that you may spoil the fabric. Therefore great care must be exercised.

The principal solvents are ether, chloroform, alcohol, turpentine, benzene, and naphtha. Each solvent may be used to best advantage on certain fabrics.

The commercial grades of the solvents often contain impurities [Pg 289] that leave a brown ring after evaporation. This brown ring is very objectionable. Turpentine is used only in removing stains from coarse fabrics. Chloroform, benzene, and naphtha are used on ordinary silks and linens. Ether and chloroform are used to best advantage in removing stains from delicate silk, as they seldom affect colors and evaporate very quickly. Of course it must be borne in mind that when a stain is removed from a fabric that portion that contained the stain loses some coloring matter and feels rougher than the other part.

Grease Spots on Heavy Goods that cannot be Laundered

It is usually desirable to use the following method in removing grease from a heavy fabric, such as carpets or colored fabrics. In case the grease is fresh, place over the stain a piece of clean blotting paper or a piece of butcher's brown wrapping paper and underneath absorbent paper or oil cloth, and then press the spot with a warm iron. As heat often affects the shades of certain colors such as blues, greens, and reds, it is best to hold a hot iron over the fabric and see if the grease is melted.

Remove a stain from a piece of carpet.

Removal of Grease and Blood

Ordinary Fabrics (wash goods). Wash the fabric containing grease or blood stain with tepid water and soap.

Delicate Fabrics. As strong soap will spoil some colors and textures it is necessary to apply a solvent when a delicate fabric is stained.

Remove stains from a washable fabric and a delicate fabric.

Removal of Wheel Grease and Lubricants on Fine Fabrics

Wheel grease is a mixture of oils and graphite. Apply benzene to the wheel grease spot. This will dissolve the oil, leaving the coloring matter (graphite) on the cloth, and this may be collected on the white cloth on the other side.

Remove a wheel grease stain from a dress fabric.

[Pg 290]

Removal of Acids

Fruit acids and all others, except nitric acid may be removed by putting ammonia on the spot. This will neutralize the acid, forming a salt which may be either brushed or washed off. In the case of nitric acid the fibers of the cloth are actually destroyed and no amount of ammonia will restore the original condition of the fabric.

Remove a stain of orange juice from a dress or shirt waist.

Removal of Blood

Blood stains may be removed from a fabric by washing with cold or tepid water. Never use hot water, as hot water coagulates the albumen of the blood. After removing the blood soap and warm water may be used. In case the fabric is a thick cloth, the blood may be removed by applications of moist starch.

Take different samples of fabrics and soil them with fruit acids, soup, wheel grease, ink, and blood and remove them. Exercise great care so as not to leave a mark or remove the coloring.

Remove blood from a fabric.

Questions

1. What is a solvent? an absorbent?

2. What is the best solvent to be used in removing stains from silks, coarse goods, and linens? from delicate silks?

3. Why is a brown ring often left after removing a stain?

4. How may grease and blood stain be removed from wash goods?

5. What is wheel grease? How may it be removed?

6. How will ammonia remove acid stains?

7. Does it remove all? Why not?

8. Explain the method of removing blood stain from cloth.

Experiment 20 — Dyeing Wool

Apparatus: Large porcelain dish or casserole, filter.
Materials: Undyed piece of woolen and worsted fabric, undyed yarn, and undyed raw cloth.
Reference: *Textiles*, page **65** .

[Pg 291]

Directions

1. Prepare a solution of coloring matter by dissolving a half ounce of diamond dye (green or red) in a quart of water. Filter the solution. Place a piece of white woolen cloth in the liquid and boil ten minutes. Then wash the dyed fabric and notice whether the dyestuff washes off or not.

2. Repeat the experiment, using the same weight of undyed woolen yarn. Repeat with worsted yarn.

3. Repeat the experiment using the same weight of wool sliver.

4. Notice which has the deeper color. The degree of color depends on the amount of twist in yarn. Which sample has absorbed the greatest amount of dyestuff from the liquid?

a. Why is a yarn-dyed fabric faster than a piece-dyed?

b. Why is a raw stock dyed fabric better than piece or yarn dyed?

Experiment 21 — Dyeing Cotton

Apparatus: Porcelain dish, filter stand, etc.
Material: Piece of cotton cloth.
Reference: *Textiles*, page **67** .

Directions

1. Prepare a solution of coloring matter by dissolving a half ounce of logwood in a quart of water. Filter the solution. Place a piece of cotton cloth in the liquid and boil ten minutes. Then wash the dyed fabric and notice whether the dyestuff washes off or not.

2. Repeat the same experiment and use a piece of cotton cloth that has been previously washed in common alum. [25] Note the effect. Which has the greater attraction for dyestuffs, cotton or wool? Why is alum used?

3. Repeat the same experiment, using first the same weight of [Pg 292] cotton yarn and then the same weight of cotton sliver. Notice the results.

Which piece of cotton holds the dye best, that which was dipped in alum or the one that was simply boiled in the solution?

Experiment 22—Weighting Silk.—Affinity of Metallic Salts for Silk

Apparatus: Porcelain dishes.
Material: Silk yarn.
Reference: *Textiles*, pages **212-214** .

Directions

1. Weigh separately two skeins of dry silk and distinguish skein No. 1 by looping some cotton thread into it. Prepare a tepid bath containing 10 gm. strong sumach extract in 400 cc. water. Enter the skeins of silk and work for 15 to 20 minutes, meanwhile slowly raising the temperature to about 150° F. Remove, squeeze, rinse with water, squeeze, and dry skein No. 1 for weighing.

2. Meanwhile prepare another bath containing 4 gm. of copperas (ferrous sulphate) in 400 cc. cold water. Work skein No. 2 in bath for 10 minutes cold. Remove, and rinse well; save the iron bath. Repeat the treatment in the sumach and iron baths several times more, finally wash the sumach iron skein in 1 per cent hot soap solution; rinse, squeeze, and dry. Weigh each dried and cooled skein and note the increase in weight of each. Save sample for Experiment **23** and note the effect of weighting on the yarn.

Experiment 23—Dyeing Silk

Apparatus: Porcelain dish, filter stand, etc.
Material: Piece of silk yarn.
Reference: *Textiles*, page **210** .

Directions

1. Prepare a solution of coloring matter by dissolving a half ounce of logwood in a quart of water. Filter the solution. Place a piece of silk skein, from Experiment **22** , in the liquid [Pg 293] and boil ten

minutes. Then wash the dyed silk and notice whether the dyestuff washes off or not.

2. Repeat the same experiment using the same weight of silk yarn without weighing it. Compare the results.

Experiment 24 — Test to Distinguish Piece-Dyed from Yarn-Dyed Fabric

Apparatus: Pen knife.
Materials: Woolen and cotton fabrics.
Reference: *Textiles*, pages **66-68** .

Directions

Unravel threads of the suspected sample, and with a blade of pen knife note whether the dyestuff has penetrated through the yarn as noted by the depth of color in the interior of the yarn. In case there is the same depth of color in the interior as on the surface, the fabric is yarn-dyed. If on the other hand, the interior of the yarn is not so highly colored as the exterior, it is piece-dyed.

Questions

1. What is meant by yarn-dyed fabric?

2. What is meant by piece-dyed fabric?

3. How may the two be distinguished?

Experiment 25 — Test to Distinguish Dyed from Printed Fabrics

Apparatus: Knife-blade.
Materials: Cotton fabrics.
Reference: *Textiles*, page **65** .

Printed fabrics may be distinguished from dyed by observing the back side of the cloth, and noting whether or not the pattern on the face of the cloth penetrates through to the back, or only the outline shows. In case the figure or pattern is on both sides of the fabric, it may be distinguished from the dyed by taking one thread of the suspected sample, and by the means of a knife-blade attempting to scrape off the coloring on the surface of the thread. If the [Pg 294] dyestuff has penetrated into the interior of the thread, it is *not* printed.

Generally speaking, printed fabrics are known from dyed fabrics by the fact that the former have the design printed on the face of the cloth. This is called *direct printing*. The best dyed fabrics are obtained by dyeing in what is called a *jig*, and the whole fabric is saturated with color. Most, if not all the cloths which you see in the retail dry goods stores which are in plain colors are dyed in the jig. Some of the cheaper qualities of dyed fabrics are padded in a mangle, but there has been a very small quantity of these goods on the market for many years.

Printed fabrics may be made as fast as dyed fabrics; it all depends upon the process by which the goods are converted. Within the past few years great headway has been made in dyeing with what are termed *vat* colors. Indanthrene is a vat color and a great many mills have used this class of dye successfully in dyeing plain shades. This is what would be termed a *fast* color in every sense of the word. There are a number of dyestuff makers in Europe who put vat colors on the market, but they all call them by different names. Vat colors have been used with success in printing during the past year or two, especially on shirting fabrics, and these colors are fast to both light and washing. Most direct colors used for printing or dyeing are equally fast to light and washing, but of course they will not stand the test as well as the vat colors mentioned above.

The essential qualities of a good printed fabric are its ability to withstand exposure to light and washing. In printing, of course, a greater variety of desirable styles can be obtained than by dyeing, in fact there are certain popular lines of goods now on the market the effect of the designs of which cannot be obtained in any other way than by printing. At the same time, although the field in designing for dyed fabrics is limited, some very handsome effects can be obtained.

It will not be many years before a large proportion of the printed and dyed fabrics put on the market, both foreign and domestic, will be in the vat colors which, as stated above, are very fast. Even at the present time there are many mills that are using this class of colors entirely, especially the mills which manufacture woven fabrics.

[Pg 295]

Questions

1. In printed fabrics is the pattern clearly discernible on the back of the cloth?

2. If the fabric is printed on both sides, how may this fact be proved?

3. What is the difference between printed and dyed fabrics?

Experiment 26 — Bleaching by Sulphur Dioxide

Apparatus: A quart bottle.
Material: Sulphur, worsted or silk fabric.

Bleaching powder cannot be used in bleaching animal fibers such as woolen and silk fabrics. It injures the fibers and at the same time leaves them yellow.

Animal fibers are best bleached by immersing in an aqueous solution of sulphurous acid or exposing them to fumes of burning sulphur.

Wet a piece of dyed worsted or silk fabric and hang it in a quart bottle containing fumes of burning sulphur. [26] The fumes of burning sulphur have an affinity for coloring matter — dyestuff. The fumes (called sulphur dioxide) do not in most cases destroy the coloring matter as chlorine does, but simply combine with it to form colorless compounds which can be destroyed. The color can be restored by exposing the bleached fabric to dilute sulphuric acid.

Questions

1. Why is it necessary that the fabric be moist in order to be bleached by sulphur dioxide fumes?

2. What becomes of the coloring matter?

Experiment 27 — Bleaching by Bleaching Powder

Apparatus: Porcelain dish.
Material: Piece of calico.
Reference: *Textiles*, page **148** .

[Pg 296]

Directions

Place a quarter of an ounce of bleaching powder in a quart bottle containing a pint of water. [27] Then place a piece of calico in the water containing the bleaching powder. What is the effect on the calico? Then remove cloth to another bottle filled with dilute hydrochloric or dilute sulphuric acid. What is the effect on the color? Then wash the whitened cloth thoroughly in water.

Why is it necessary in practice to pass cotton fabrics through two baths in bleaching? What is contained in the first bath? in the second bath?

Experiment 30 — Determining Style of Weave

Apparatus: Pick glass.
Materials: Different fabrics.
References: *Textiles*, pages **56-58** , etc.

Examine different samples of cloth and classify them according to the seven standards given on pages **56-58** , etc.

Experiment 31 — Determining the Size of Yarn

Apparatus: Yard stick.
Materials: Sample of cotton, woolen, and worsted yarns.
References: *Textiles*, pages **49** , **51** , **52** .

As yarns used in the manufacture of fabrics are of all degrees of thickness, it became necessary to adopt some method of measuring this thickness. For this purpose yarns are numbered, so that when the number is known an idea of the size of the yarn may be gained. It would seem advisable to number yarns of all kinds [Pg 297] according to one fixed standard, yet unfortunately this is not done. The methods of counting yarns are many and varied. The usual method is to estimate the yarn number by taking the number of hands of a definite length which make up some given weight. Thus in the worsted yarn, No. 1 is a yarn that has 560 yards to a pound. No. 2 worsted yarn has two times 560 yards to a pound. How many yards in No. 12 worsted yarn? How many yards in No. 20 cotton yarn?

Experiment 32 — Test for Twist in Yarn

Apparatus: Test dial.
Material: Piece of yarn.
References: *Textiles*, pages **131-132** .

As the amount of twist in yarn determines its strength, it is necessary to know the amount of twist per inch in given yarn. The strength increases up to a certain limit. When this limit is reached, increased twist does not make the thread any stronger. We may also have twist and strength at the expense of bulk. The test consists in finding out the number of turns per inch, and this is done by an arrangement where a certain length of yarn is stretched between two points on a twisting machine and the twist taken out. The number of turns required to take the twist completely out are registered on a dial at the side of the apparatus.

Poor cotton that goes into coarse goods cannot be spun as fine as the finer cotton. The shorter the cotton the more twist is required to spin it, and the more twist that is put into the yarn, the less will be the yardage. Whereas on the finer and longer cotton there will be less twist put into it, and the yarn will be much stronger. Find the twist in different kinds of yarn.

Experiment 33 — Determining the Direction of Warp and Filling

Apparatus: Microscope.
Materials: Silk, cotton, and woolen fabrics.
Reference: *Textiles*, page **238** .

When one examines a fabric the first thing to do is to determine the direction of the warp and direction of the filling.

[Pg 298] Fabrics with Selvedge.—Examine any fabric with a selvedge and notice that the warp threads run in the same direction as the length (longest side) of the selvedge. What direction will the filling threads bear to the selvedge?

Fabrics with a Nap.—Examine a piece of flannel and notice the direction of the nap. Why will the direction of the nap be the same as the direction of the warp? Remember the way in which the fabric enters the napping machine.

Fabrics Containing Double Threads.—Examine a fabric containing double and single threads and notice that the warp contains the double threads. Why?

Fabrics Containing Cotton and Woolen Yarn.—Examine a fabric containing cotton and woolen threads running in different directions and notice that the cotton threads form the warp. Why?

Another way to tell the warp threads in a fabric is to examine warp and filling threads very closely and notice which set contains the greater twist? Why? See if they are separated at more regular intervals. Why?

Stiffened or Starched Fabrics.—Examine stiffened or starched goods very closely and notice the threads. If only one set can be seen they are the warp threads. The stiffer and straighter threads are found in the warp. Why? The rough and crooked threads are seen in the filling.

Experiment 34—Determining the Density of a Fabric

Apparatus: Pick glass.
Materials: Samples of cloth.
Reference: *Textiles*, page **238** .

Directions

1. Examine different samples of cloth and determine the number of filling threads and warp threads by means of a pick glass.

2. Then examine different priced fabrics of the same kind and see whether the low or high priced fabric has the greater density.

[Pg 299]

Experiment 35 — Determining Weight

Apparatus: Balances, die.
Materials: 4 square inches of cloth.
Reference: *Textiles*, page **239** .

Directions

1. Fabrics are bought and sold by the yard. In order to express the amount of wool or cotton in a fabric the weight in ounces per yard is usually given.

2. In order to find the number of ounces per yard a piece of cloth of definite size, usually about 4 sq. in., is stamped out by means of a die, or cut by means of a tin plate exactly 4 sq. in. (2 in. on the side). This is then weighed on very accurate balances and expressed in grains. Find the weight per yard. Remember 7000 grains equal 1 lb.; 16 oz. equal 1 lb.

Experiment 36 — Determining Shrinkage

Apparatus: Hot water.
Material: Sample of woolen fabric.
Reference: *Textiles*, page **239** .

Directions

1. Take a sample of a woolen fabric 12 in. by 20 in. and pour hot water over it and leave it immersed over night. Then dry it in the morning at a moderate temperature without stretching. Then measure its length and divide the difference in lengths by the original length. The quotient multiplied by 100 will give the per cent of shrinkage.

2. Repeat the same experiment with a worsted fabric, and with a cotton fabric.

3. Why does the woolen fabric shrink more than the worsted?

Experiment 37 — Test of Fastness [28] of Color under Washing

Apparatus: Porcelain dish, soap solution.
Materials: Cotton and woolen fabrics.
Reference: *Textiles*, page **242** .

[Pg 300]

Directions

1. Colored goods and printed fabrics should withstand the action of washing. They require more care than white goods and should be soaked in cold water containing very little soap and no soda. They should be dried in the shade as a very hot sun will fade them. If it is necessary to dry them in the sun be sure that they are dried wrong side out, as direct sunlight fades them about five times as much as reflected light.

2. All colored fabrics should stand mechanical friction as well as the action of soap liquor and the temperature of the washing operation. In order to test the fabric for fastness a piece should be placed in a soap solution similar to that used in the ordinary household, and heated to 131° F. The treatment should be repeated several times.

3. If the color fails to run it is fast to washing.

Questions

1. Why should more care be exercised in washing colored goods than white?

2. How may colored fabrics be tested to show that they stand the action of soap solution?

3. Does a moderately warm temperature (131° F.) affect the fastness of colored fabric?

Experiment 38 — Test of Fastness of Color under Friction

Apparatus: Yarn, white unstarched cotton fabric.
Materials: Fabrics worn near the skin.
Reference: *Textiles*, page **242** .

Directions

Stockings, hosiery yarns, corset stuffs, and all fabrics intended to be worn next to the skin must be closely knitted to withstand friction and must not rub off, stain, or run, that is, the dyed materials must not give off their color when worn next to the human epidermis (skin), or in close contact with other articles of clothing, as in the case of underwear.

In order to test two fabrics to see which is the better, it is [Pg 301] necessary to rub the fabric or yarn on white unstarched cotton fabric.

In comparing the fastness of color of two fabrics it is necessary to have the rubbing equal in all cases.

Questions

1. What is meant by friction?

2. What is meant by the expression "fastness of color of two fabrics"?

3. How may the fastness of a colored material be tested to withstand friction?

Experiment 39 — Test of Fastness of Color against Rain

Apparatus: Water, undyed yarn.
Materials: Silk and woolen fabrics.
Reference: *Textiles*, page **243** .

Directions

Silk and woolen materials for umbrella making, raincoats, etc., are expected to be rainproof. These fabrics are tested by plaiting with undyed yarns and left to stand all night in water. Notice whether the color of the fabric has run into the undyed yarns.

Take a sample of the fabric and shake some drops of water on it. Notice whether it loses its luster when the drops have dried. Spotting may be prevented by placing a damp cloth on the wrong side of the material; roll the two together, and when evenly damp, unroll and press through the damp cloth with a fairly hot iron.

Place a piece of the fabric in the sun so that the sun and rain may come in contact with it. Notice whether it loses its color and becomes gray and dull.

Experiment 40 — Test of Fastness of Color in Sunlight

Apparatus: Cardboard.
Materials: Silk, woolen, and cotton fabrics.
Reference: *Textiles*, page **244** .

[Pg 302]

Directions

Cover one end of the sample of cloth with a piece of cardboard. Expose the fabric to the sunlight for a number of days and examine the cloth each day and notice whether the part exposed has changed in color when compared with the part covered. Count the number of days it has taken the sunlight to change the color. Does direct sunlight have any effect upon colored fabrics? Which is the most affected by the sun, silk, woolen, or cotton fabrics, dyed with same dyestuff, in the same length of time? Are fabrics changed any sooner by the sun than by the weather?

Experiment 41 — Test of Fastness of Color to Weather, Light, and Air

Materials: Cotton, silk, and woolen fabrics.
Reference: *Textiles*, page **244** .

Directions

Examine various fabrics for fastness to weather, light, and air by placing samples outside of a window so that they will be exposed to the weather, light, and air. Have duplicate samples of the above away from the weather and light. Compare the samples exposed to the weather with those in the house and note the number of days it takes to change. Classify the fabrics. Which of the fabrics are most easily affected by the weather, light, and air?

Experiment 42 — Test of Fastness of Color against Street Mud and Dust

Apparatus: Porcelain dish, lime, and water.
Materials: Cotton, silk, and woolen fabrics.
Reference: *Textiles*, page **243** .

Ladies' dress goods are expected to withstand the action of mud and dust. In order to test a fabric for the resistance, the sample should be moistened with lime and water (10 per cent solution), dried, and brushed. Or sprinkle with a 10 per cent solution of soda, drying, brushing, and noting any change in the color.

[Pg 303] *a.* Is there any change in color after the lime water has been removed from the cloth?

b. With what may the action of the lime water or soda be compared?

Experiment 43 — Testing Rubberized Fabrics

Materials: Rubberized fabrics.

A great many rubberized fabrics are used for hospitals, domestic purposes and for clothing. On account of the high price of excellent rubberized fabrics a great many substitutes are placed on the market that are satisfactory to the eye, but have not the wearing qualities for the service they are intended to render.

Strength and Resistance to Scratching. — In order to test a rubberized fabric to see if it has the necessary strength to stand everyday use, see if it is possible to scratch it with the finger nail. Then crease it and crumple it between the hands. Then spread it out very carefully and notice whether there are any broken places. If there are it should be rejected.

Waterproof Qualities.—A rubberized fabric should be waterproof. A sample may be tested by forming a bag with it and filling it with water. Crumple the bag while it is filled with water. Notice whether it cracks or leaks.

Examine various rubberized fabrics and notice whether they are substantial.

Experiment 44—Test for Vegetable and Animal Fiber

Apparatus: Acid, fire.
Materials: Warp and filling threads of cotton and woolen fabrics.
Reference: *Textiles*, page **239** .

Directions

One of the most useful tests is to see whether an article is made of wool, cotton, or silk, and if a composition of two or more materials, to estimate the percentage of each. Practical experience can teach one much in this respect, and in many cases inspection is quite insufficient. A more reliable test is to burn a piece of material and notice how it burns.

[Pg 304] Take a sample of a woolen and cotton fabric; separate the warp and filling and untwist one piece of warp and one piece of filling yarn. Burn a piece of untwisted yarn and notice whether it burns slowly and curls up into a black crisp cinder leaving a disagreeable smell, or burns with a flash leaving a light ash behind.

Questions

1. Describe the burning process.

2. What is the burning test for vegetable fiber?

3. What is the burning test for animal fiber?

Repeat the same experiment, placing the untwisted yarn in sulphuric acid. Apply heat and note the effect.

What is the acid test for vegetable fiber?

What is the acid test for animal fiber?

Examine different fabrics to see whether they contain vegetable or animal fibers.

Experiment 45 — Difference between Cotton and Linen Fabrics

Examine a real linen towel and a cotton towel. Wet your hands and use both towels to dry them. Notice which of the fabrics absorbs the moisture quicker, or which towel dries the hands better.

Compare a cotton table-cloth and a linen table-cloth. Notice that the linen fabric has a natural gloss, a cool, smooth feel, and launders much better than cotton. The cotton fabric on the other hand gives off a fuzz, and irons dull and shapeless. [29]

Linen is tough and strong, cool feeling, and has a long fiber. Linen cannot be given a cotton fabric finish. [30]

[Pg 305] Cotton on the other hand has a weak, short fiber, dull, warm, and non-absorbent. After washing, cotton resembles a limp rag while linen retains firmness and stiffness.

Which fabric absorbs the moisture more readily?

What is the difference in appearance between the two fabrics? Between the fibers of the fabrics?

Experiment 46 — Test to Distinguish Artificial Silk from Silk

Apparatus: Porcelain dish, potassium hydrate.
Material: Piece of silk fabric.
Reference: *Textiles*, page **240** .

Since silk fabrics, particularly hosiery, are becoming popular, various attempts have been made to produce substitutes for real silk. To test a silk fabric, boil the sample in 4 per cent potassium hydrate solution and note the effect. If it produces a yellow solution it is artificial silk, if colorless it is pure silk.

Another simple way used by some workmen, although unhygienic, is to unravel a few threads of the suspected fabric, place them in the mouth, and masticate them vigorously. Artificial silk will soften under the operation and break up into a mass of pulp. Natural silk will retain its fibrous strength.

Test various samples of cheap "silk" hosiery.

Experiment 47 — Test to Distinguish Silk from Wool

Apparatus: Porcelain dish, hydrochloric acid.
Material: Silk or woolen fabric.
Reference: *Textiles*, page **240** .

[Pg 306] Silk may be distinguished from wool by putting the suspected thread or fabric into cold concentrated hydrochloric acid. If silk is present it will dissolve, while wool merely swells.

Test various samples of silk and wool.

Experiment 48 — Test to Distinguish Cotton from Linen

Apparatus: Fuchsine, ammonia.
Materials: Cotton and linen fabrics.
Reference: *Textiles*, page **240** .

Directions

1. On account of the high price of linen various attempts are made to pass cotton off for linen. While it is possible sometimes to detect cotton by rolling the suspected fabric between the thumb and finger, the better way is to stain the fabric with fuchsine. If the fibers of the fabric turn red, and this coloration disappears upon the addition of ammonia, they are cotton; if the red color remains, the fibers are linen. The most reliable test is to examine the fiber under the microscope and note the difference in structure.

2. Test a cheap *linen* fabric for cotton.

3. When cotton yarn is used to adulterate linen it becomes fuzzy through wear, and when used to adulterate other fabrics, it wears shabby and loses its brightness. Linen is a heavier fabric, and wrinkles much more readily than cotton. It wears better, and has an exquisite freshness that is not found in cotton fabrics.

4. Describe a chemical test for linen.

5. What is the difference in appearance and wearing qualities of cotton and linen?

Experiment 49 — Test of Fabric to Withstand Ironing and Pressing

Apparatus: Hot iron.
Materials: Silk, cotton, and woolen fabrics.

Directions

Place the sample over an ironing board and iron it with hot iron (about 200° F.). Compare the sample immediately with one [Pg 307] not ironed. Remember that many colors, particularly colored silks, change while they are hot. If the original shade returns when the fabric has cooled, then the fabric is fast to ironing and pressing.

Questions

1. What is the object of ironing and pressing clothes?

2. Is it a good plan to press clothes often?

3. Is there any difference in the effect of the hot iron on the three kinds of fabrics?

Experiment 50 — Test of Fabric to Withstand Perspiration

Apparatus: Porcelain dish.
Materials: Silk, cotton, and woolen fabrics.
Reference: *Textiles*, page **243** .

Directions

1. In addition to withstanding the action of coming in contact with the human skin, fabrics like hosiery, etc., should withstand the excretions of the body.

2. To test a fabric for resistance, place the sample in a bath of dilute acid made by adding one teaspoonful of acetic acid to a quart of water warmed to the temperature of the body, 98.6° F. The fabric should be dipped a number of times, and then dried, without rinsing, between parchment paper.

Questions

1. What is the effect of the acid solution upon the fabrics?

2. Would they necessarily withstand the effect of perspiration, even if they did withstand the acetic acid solution? Why?

Experiment 51 — Test for Determining Dressing

Apparatus: Magnifying glass, porcelain dish.
Materials: Various fabrics.
Reference: *Textiles*, page **242** .

Directions

1. A great many cotton fabrics such as muslin often contain considerable sizing or dressing. In order to examine a fabric [Pg 308] and determine whether too much dressing has been used, take a small sample of the fabric and crush it in the hand and rub it together, so that the dressing is removed and the quantity employed may be determined. If much dressing has been used, dust will be produced in rubbing. Prick the surface with your finger nail. Notice whether the starch comes off. Then wet your finger and rub it on the cloth and allow it to dry. Does the gloss disappear?

2. Another method is to hold the sample before the light and notice whether you can recognize the dressing. Examine the sample with a magnifying glass (or pick glass) and notice whether the dressing is superficial or penetrates the substance of the fabric.

3. Would you buy low priced cotton goods with a thick gloss and pasty look?

4. Notice the effect (lusterless) of fabrics containing much dressing after washing.

5. A very simple way for telling the amount of loading or weighting in a cotton fabric is to weigh a given sample, then "boil the fabric out in hot water," —boiling for several hours and then drying it. The difference in weight after drying and before boiling gives the weight of sizing material per sample piece.

6. If mineral loading has been used to a great extent, a large residue is left after burning.

Experiment 52—Testing the Strength of Cloth

Apparatus: Dynamometer.
Materials: Various fabrics.
Reference: *Textiles*, page **237** .

Directions

1. An excellent way to test the strength of a fabric is to place the two thumbs together and press them down on the sample, holding it tight underneath. Then try to break the threads, first in one direction and then in the other. Do they break easily? Notice whether one set is very much stronger than the other.

2. Manufacturers usually test yarn and fabrics by means of an instrument called a dynamometer. In this way one can find out whether a yarn or fabric comes up to the necessary strength, and [Pg 309] whether it has the required yield or stretch. Both these points are of importance in practical work, for it is essential that the yarn as shown should at least be strong enough to bear the strain of manufacture. The test is made by stretching a hank of yarn between the two hooks of a cloth testing machine. The handle at the side is now turned, so that the lower hook descends and puts a strain on the hank. This strain is increased, and at the same time the pointer moves around the dial, which indicates in pounds the amount of strain. When the threads of the hank begin to break, the strain is

released, and the catch at the side keeps the pointer in position until the amount of strain is read on the dial. The distance stretched by the yarn before breakage occurs is shown in inches and fractions of an inch, in the small indicator arranged near the upper hook.

Test different fabrics and yarns.

Experiment 53—Characteristics of a Knitted Fabric

Apparatus: Pick glass.
Materials: Knitted fabric, woven fabric.
Reference: *Textiles*, page **153** .

Directions

1. Examine a piece of knitted fabric under the pick glass and notice the construction. How does it differ from weaving? The single thread is formed into rows of loops which hang upon each other, thus giving the knitted fabric its characteristic springiness. Why is hosiery suitable for underwear? Try to obtain the thread of the knitted fabric and reduce the whole to a heap of yarn by cutting it. Is the yarn intact?

2. A knitted fabric may be told from a woven fabric by studying the following sketch. (See page **310** .)

Note that the element of stretch or elasticity is wholly lacking in the woven cloth except what lengthwise elasticity may be in the threads themselves. On the other hand, referring to the printed diagram of the knitted fabric it will just as readily be seen that its very structure implies such a corrugation of its individual loops that if distended by force in any direction its tendency is to return to the normal.

[Pg 310] The essential characteristics of good hose are:

1. That they should be without seams.

2. That they should be so knit as to conform to the foot of the wearer.

3. That they should be thickened or reinforced where the greatest wear comes.

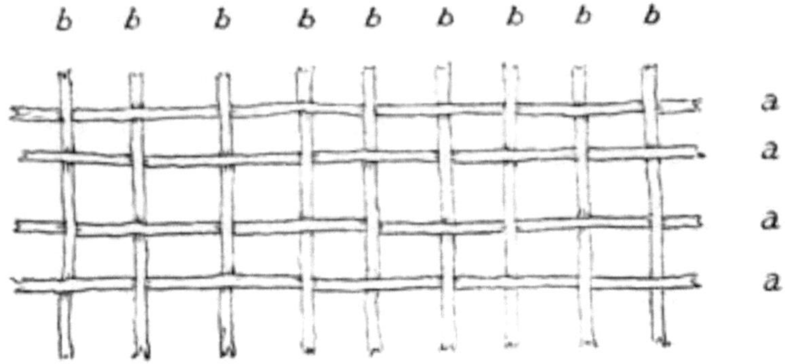

ORDINARY WEAVING
a. Weft. *b.* Warp.

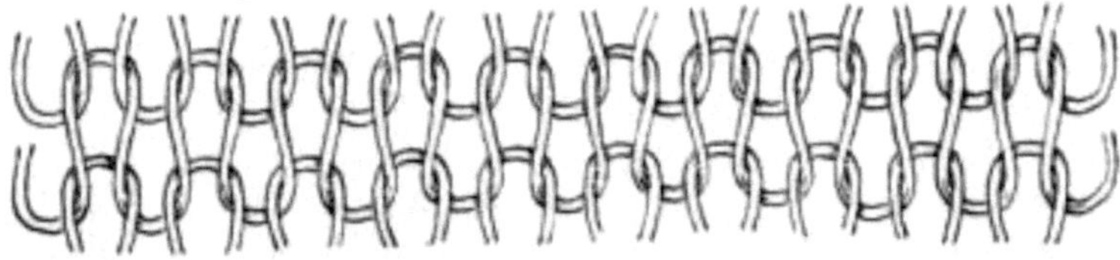

PLAIN STOCKING FABRIC

The essential characteristics of good underwear are:

1. It should be made from elastic cloth, which implies a knitted cloth.

2. It should be porous in a high degree—whether in winter weight or summer weight.

3. The seams should be made upon the most modern sewing machines, with as little bulk as possible.

Experiment 54—Experiment to Illustrate Imperfections in Hosiery

Apparatus: Magnifying glass.
Materials: Hosiery (seconds).
Reference: *Textiles*, page **159** .

[Pg 311] Examine a piece of hosiery called a *second*, obtained from a hosiery mill, and notice whether any of the following defects are present:

a. Yarn contains thick bits and buns.

b. Knots are badly tied.

c. Discolored yarn.

d. Lacks solidity and firmness, due to the gauge being too coarse for the yarn.

e. Full work-yarn too thick for gauge.

Experiment 55 — Characteristics of a Crochet Fabric

Apparatus: Crochet needle.
Material: Thread.

Directions

1. Crocheting is another method, like knitting, of making a fabric. Examine the operation of crocheting. Notice that only one thread is made use of, and is formed into loops by means of a large needle with a hook at the end. The chief point is that the loops are not formed in uniform rows, but one loop at a time, and with the greatest liberty to choose any part of the article already made to form a new loop. For this reason, crocheting adapts itself to the production of fancy patterns useful for ornamenting.

2. A knitted fabric can be told from a crocheted fabric by noting that a knitted fabric is composed of rows or ranks of loops of a single yarn which interlock successively with similar rows or ranks of loops, while a crocheted fabric consists of a structure the basis of which is a thread crocheted or knitted into a chain which is attached at intervals to itself.

Experiment 56—Characteristics of a Good Piece of Cotton Cloth

Materials: Different samples of cotton fabrics.
Reference: *Textiles*, pages **151-152** .

Directions

1. Examine different kinds of cotton fabrics. Compare the lowest, medium, and highest priced varieties of the same fabric. [Pg 312] Notice how quickly the finish of a low priced cotton fabric with a *thick* gloss loses its luster after washing.

2. Examine the different grades of the following fabrics: white lawn, Indian Head cotton suiting, muslin, lawn, and cambric. Wash the samples several times.

Does it pay to buy cheap cotton fabrics for underwear, etc.?

What are the characteristics of a good piece of cotton cloth?

Experiment 57—Characteristics of a Good Piece of Woolen Cloth

Materials: Different samples of woolens.
Reference: *Textiles*, pages **71-82** .

Directions

1. Woolens differ from worsteds in having a more or less covered face, with the result that the weave rarely is noticeable, and the general color effects are much smoother and softer than those of worsteds.

2. Examine different grades of woolen fabrics, such as serges.

Defects.—If a piece of woolen is not constructed right from the start or if the work is not properly finished, that is, enough fulled in width or length, it is liable to be raggy or slazy. As a great many fabrics are more or less teaseled, there is a possibility of such pieces becoming too woolly and too hairy.

Experiment 58 — Characteristics of a Good Piece of Worsted Cloth

Materials: Different kinds of worsted fabrics.
Reference: *Textiles*, pages **71-82** .

Directions

1. A good piece of worsted fabric should have a clear outline of the pattern, perfection of weave lines, and when the fabric is exposed to light should show a luster without polish.

2. Examine different types of worsted fabrics, and notice how many conform to the above requirements.

3. The most essential point of worsted is that it should have a clean [Pg 313] and even looking face. By clean is meant well sheared. By even is meant that the individual ends and picks used should be *even* and not full of knots, or of any foreign matter. Of course, there are some exceptions, for instance, in an unfinished worsted which has more or less nap on the face, it could not be sheared absolutely clear, but at the same time, the face should be very evenly cropped.

Defects. — A serious defect would be if the cloth was not well sheared or if it contained many uneven cords and picks, or ends and picks missing, or coarse ends and slubs.

Examine different worsteds and notice any defects.

Worsteds. — Speaking generally, worsteds may be divided into two classes, distinguishable according to the *luster* of their surface, or to the softness of their feel. They are used both for ladies' and men's wear. Worsted coatings may also be classed as worsteds. The coatings are woven in both single and double cloths in fancy weave effects for piece dyes, marketed in variety of finish, according to fashion.

Under this heading may be classified staple cloths, such as serges, clays, and fancy weave effects without any illumination. They can be finished in three ways: — Clear, undressed, and cheviot, used for ladies' dress goods or men's wear, according to weight.

The finish of the cloth varies according to the fashion, but there is always a certain demand for clear and undressed worsteds, for men's wear.

Examine a number of worsted fabrics and classify them.

Suitings.—The term suitings covers various manipulations of manufactured goods.

1. Tennis suitings, composed of all wool, or all worsted, white or cream ground, decorated with solid color, silk and weave stripe effects.

2. Piece-dyed worsteds, such as a blue ground with white silk line, cable cord, and fancy weave stripe effects, or any other ground shade color with its complementary decoration applied.

3. Mixture wool or mixture worsted yarns made into fabrics, decorations applied in color; cable, silk, and weave effects in stripes or overline color checks, suitable for men's wear, or decorated suitable for woman's wear. The darker shades for fall and the lighter shades for spring.

[Pg 314] General weight of fabric for men's wear, 12 to 14 oz. per yd., 56 in.; general weight of fabric for ladies' wear, 8 to 12 oz. per yd., 54 in.

4. As a rule, when one speaks of a suiting, you expect to see a fancy effect, in the form of a fancy stripe, check, or a colored mixture, in loud or quiet tones of decoration. Long naps in fancy effects are sometimes fashionable, and at other times the cloth finish is popular.

This class may be subdivided into

1. Light weight for spring or fall.

2. Heavy weight for winter.

The light weight class generally consists of covert cloths in lighter colors for spring, and cloths usually of the undressed finish from worsted or woolen stock for fall.

The heavy weight class generally consists of heavily fulled goods, such as meltons, beavers, naps, etc., which give a heavier and warmer coat for winter use only, and where an exceptionally heavy coat is required, double and treble cloths are occasionally employed.

Examine different kinds of suitings and classify them.

Trouserings. — Trouserings are more firmly woven than suitings and are heavier. They invariably have a stripe. The ground shade of the better grade of men's wear fabrics is generally composed of twist warp yarns, ranging from dark slate gray to light lavender gray. An endless variety of broad and narrow fine line effects is produced by expert manipulation and combination of weave and silk decorations, producing the pleasing effect required for this class of goods. The filling is nearly always black; but sometimes a dark slate is used.

The cheaper grades are generally made of wool and cotton mixtures and twists, down to all cotton, in imitation of the better grades.

Overcoatings. — Overcoatings are heavy woolen or worsted fabrics and heavily teaseled or gigged, giving a rough, hairy appearance. Whether thick or thin, coarse or fine, they should always be elastic fabrics, that is, as much so as well fulled woolen goods can be. When hard or stiff they never make a graceful garment. The special goods made for overcoats are nearly all soft goods.

Examine different fabrics and classify them into either suitings, overcoatings, trouserings, etc.

[Pg 315]

Experiment 59 — Characteristics of a Good Silk Fabric

Materials: Samples of different cheap silk fabrics.
Reference: *Textiles*, pages **203-218** .

There are cheap and expensive silk fabrics on the market. The consumer is often tempted to buy the cheaper fabric and wonders why there is such a difference in price. The difference in price is due to the cost of raw material and additional cost is due to the care in manufacturing. For example, raw silk costs from $1.35 to $5 a pound according to its nature, quality, and the country from which it comes. The cost of throwing silks preparatory to dyeing also varies, the average being 55 cents a pound for organzine or warp, and 33 cents a pound for tram and filling. The prices here also vary according to the nature of the twist imparted to the silk, which is regulated by the kind of cloth it is to enter into. The cost of dyeing varies from 55 cents a pound upwards to perhaps $1.50 a pound, according to the dye and the treatment which the silk is to receive in the process of dyeing. The cost of winding, quilling, and sundry labor items necessary with soft silk prior to its being woven, will perhaps average about a cent per yard of woven goods for the cheapest cloths and range upwards according to the grade of the fabric. The cost of weaving also varies with the cloth, and may be 9 cents for one fabric and 25 cents or more per yard for the more expensive.

Weavers are paid from 2 cents to 60 cents per yard for weaving the different fabrics, and other operations vary greatly in cost; for instance, the cost of printing is entirely dependent upon the work and the number of colors used, whether it is blotch printing, discharge work, or block printing. Different processes in finishing have widely varied costs. At the present time moire work is done which costs as high as 25 cents per yard. There are also other materials which can be finished for as little as ½ cent per yard. Some goods have to be finished over and over again in the dyeing and finishing while others are very simply done. Many printed goods are handled 150 times after they come from the loom.

When it comes to relative values of similar goods produced by different manufacturers there are a few general principles by which [Pg 316] good construction can easily be determined. Most pure dye fabrics when burned will rather shrivel and boil than burn, while those which are weighted heavily with metallic salts will simply char and turn white without losing the structure of the fabric.

A fabric in which the quantity of warp and filling are of equal weight gives the maximum strength for the amount of material used. For the same weight and material, that having the most bindings of warp and filling will give the greater service. Fabrics with an insufficient number of warp or filling threads slide easily and do not give good service, though sometimes fashionable. A fabric having a twist in the warp and filling will last longer than one using the same amount of silk and the same binding with less twist.

All of these things may be taught to women many times over, but if the fashion demands an article which breaks all of the above laws and is everything that it should not be, they will buy it in preference to a serviceable fabric. As a general rule, the consumer will be safest in buying goods produced by houses of good reputation and whose products are well known.

A large part of the retail value of silk goods is their fashion demand and is quite independent of their cost of production. For instance, at the present time crêpe fabrics, brocades, and prints are commanding a premium while such goods as plain taffetas could not be sold for the cost of production.

The advantages of the better kinds of silks over the cheap ones are pure dye, long wear, and more expensive manufacturing.

Experiment 60 — How to Determine the Count of Yarn in Cloth

Apparatus: Scales, ruler.
Material: Samples of fabrics.
Reference: *Textiles*, pages **144-146** .

The United States Government imposes a tax on certain imported fabrics and yarn. In the case of cotton, the rates of duty are to be ascertained according to the average number of the yarns in the condition in which it is imported.

The length of the yarn is to be counted as equal to the distance [Pg 317] covered by it in the cloth, all clipped threads to be measured as if continuous and all ply yarns to be separated into singles and the count taken of the total singles; any excessive sizing is to be removed by boiling or other suitable processes. The number of the yarn is the English number of 840 yd. to a lb. for a No. 1 yarn.

The average number of yarn may be found without unraveling the fabric, and is the quotient of the total thread length, by the weight in the proportion of 840 yd. of yarn equaling 81/3 grains, which is equivalent to a No. 1 yarn.

The following simple formula may be used:

Multiply the count of threads per square inch by the number of square inches in the sample used, this product to be multiplied by 100; then divide the product thus obtained by the weight of the sample in grains multiplied by 432. The quotient will give the number of the yarn.

For example, take a sample of cotton cloth 4 in. square, which equals 16 sq. in., having 28 warp and 28 woof threads, a total of 56 threads to the square inch, and weighing 8.6 grains. The formula applied would be as follows:

$56 \times 16 \times 100 \div 8.6 \times 432 = 24$, the number of the yarn.

The formula may be further simplified by weighing a square yard of the cloth and dividing the number of threads per square inch by 1/300 of the weight per square yard in grains.

Find the number of yarn in several cotton fabrics.

Experiment 61 — Study of Fabrics

A great deal of time should be devoted to the study of standard fabrics so that pupils may be able to recognize them by inspection and know how to test them for adulterants.

This may be done by having the pupils study the fabrics one by one, placing a sample of each in a note-book. Underneath the sample should be written the use of the fabric, the width, the different grades, with prices, wearing qualities, and how the fabric is made. In connection with this work special effort should be made to develop a textile vocabulary so as to be able to discriminate between the different fabrics, to know the types of weaves, and the different kinds of finish, etc. In this [Pg 318] way develop the ability to know what materials and colors weave best, the prices which should be paid for strong materials, the amount of material necessary, and the trade names of fabrics which can be depended upon for substantial goods.

Occasional tests in recognizing fabrics should be given by the teacher by placing before the pupils unlabeled fabrics that they have previously studied and have them give the name, approximate price or grade, weave, qualities, etc.

Remnants or small pieces of standard fabrics may be obtained from the leading dry goods stores of the country. Teachers should have on exhibition in cabinets a large display of standard fabrics with a card attached giving the name and use of each.

Experiment 62 — How to Examine a Fabric

The first thing a buyer of cloth notices in examining the fabrics is the finish. The finish is tested by feeling and seeing. To illustrate: broadcloth should have a smooth face and a nap evenly laid. If the finish is in keeping with the character of the cloth, he next examines the fiber of the yarn to see whether it is composed of pure wool or two or more fibers in combination.

Then test the strength of the cloth to see if it will meet the requirements.

A test should be made to tell whether it is poorly dyed or well dyed. There is no test that can be applied to all colors to ascertain this, neither is it possible to judge by the eye. The best way is to take a small sample of the goods and submit it to the washing and light test.

FOOTNOTES:

[21] Dissecting pin may be made by placing head of pin or needle in a pen holder.

[22] A hand loom consisting of simply a square frame, may be obtained from Hammett & Co., Devonshire Street, Boston, Mass.

[23] In the case of linen the short fibers separated by combing are called *tow*, and the long fibers *line*.

[24] Absorbents are substances that will absorb readily excess of liquids; they include varieties of chalk, paste of chalk, or fullers' earth, rough surface of a visiting card, buckwheat flour, crumbs of bread, powdered soapstone, pumice, whiting. These substances are used to great advantage in assisting to remove stains from delicate

fabrics. They absorb the excess of solvent and thus prevent it from spreading.

[25] Alum in this case is called a mordant, which is a substance that will impregnate the cloth with something which will hold the coloring matter. Other mordants are oxides, hydroxides, and basic salts of aluminum, iron, tin, and chromium.

[26] Place a piece of sulphur on a deflagrating spoon and light it by placing it in the flame and allow it to burn. Cover the bottle by means of a glass plate.

[27] Bleaching powder is prepared by passing chlorine gas over layers of slaked lime (lime to which a slight amount of water has been added). Bleaching powder bleaches by having its hypochlorous acid set free, which in turn gives up oxygen, being converted into hydrochloric acid. The French use solutions containing chloride and hypochlorite of soda. They are called Labarraque's disinfecting fluid. A similar solution of a mixture of chloride and hypochlorite is called Eau de Javelle.

[28] A description of shoe and hand clothing may be obtained from *Shoemaking*, published by Little, Brown & Co., Boston.

[29] In Ireland the cost of producing a pound of bleached linen cloth 4 sq. yd. is 16*d.* or 32 cts.; cost of hackling a pound of flax is ½*d.* or 1 ct. per lb.; cost of preparing and spinning a pound of flax is 6*d.* or 12 cts. per lb.; cost of winding and weaving a pound of flax is 2½*d.* or 5 cts. per lb.; cost of bleaching and finishing a pound of flax is 7*d.* or 14 cts. per lb.; $75 is spent in turning $100 worth of flax into yarn; $75 is spent in turning $100 worth of yarn into brown linen; $50 is spent in turning $100 worth of brown linen into linen for market.

[30] A linen fabric can be best told from cotton by holding it up to the light and examining the evenness of the threads. Cotton can be more easily spun level than flax, therefore threads that present considerable irregularities may be taken to be flax. In a union fabric the nap is usually cotton and the threads more regular than the filling (flax). The best linen is made from fine and fairly regular threads; common linen from coarse and irregular tow yarns. Linen is no more subject to weak places in weaving than cotton, although it is

harder to bleach and may be weakened in this process. If each operation is not perfect the linen will become yellow in storage.

[Pg 319]

SOURCES OF SUPPLY

The author has found that very nearly all manufacturers are willing to supply schools with samples of their products. But the demand for samples has been so great that it is necessary in most cases to pay a small sum to cover the cost.

The following prominent firms dealing in textile supplies are named here to assist the teachers in writing for supplies.

The names of the leading textile papers are given so that teachers may obtain them. They contain a large number of names of dealers in textiles so that they may be used as reference books for supplies.

Catalogues of Cotton Machinery

Kitson Machine Shop, Lowell, Mass.—Cotton pickers.

Howard and Bullough, Pawtucket, R. I.—Cotton machinery.

Saco-Pettee Machine Shop, Saco, Me.—Cotton machinery.

Lowell Machine Shop, Lowell, Mass.—Cotton machinery.

Whitin Machine Works, Whitinsville, Mass.—Cotton machinery.

Mason Machine Works, Taunton, Mass.—Cotton machinery.

Draper Co., Hopedale, Mass.—Cotton machinery.

Woonsocket Machine Works, Woonsocket, R. I.—Cotton machinery.

Faler & Jencks, Pawtucket, R. I.—Cotton machinery.

Potter & Johnson, Pawtucket, R. I.—Cotton machinery.

C. E. Riley, 65 Franklin St., Boston, Mass.—Cotton machinery.

Cohoes Iron Foundry Co., Cohoes, N. Y.—Cotton machinery.

American Moistening Co., 120 Franklin St., Boston, Mass.—Humidifiers and textile machinery.

Standard Textile Papers

American Wool and Cotton Reporter, Atlantic Ave., Boston, Mass.

American Silk Journal, East 28th St., New York City, N. Y.

[Pg 320] Textile World Record, Congress St., Boston, Mass.

Technical Education Bulletin on Illustrative and Laboratory Supplies, published by Teachers College, Columbia University, West 120th St., New York.

Fibre and Fabric, 127 Federal St., Boston, Mass.

Textile Manufacturers Journal, Atlantic Ave., Boston, Mass.

Wool, Cotton, and Silk Samples

American Woolen Co., Boston, Mass.—Booklets on *From Wool to Cloth*; samples of fabrics.

Arlington Mills, Chauncey St., Boston, Mass.—Samples of cotton and wool in different stages of manufacture; fabrics.

S. Blaisdell, Jr., Chicopee, Mass.—Egyptian and Peruvian cotton, etc.

Frank A. Tierney, 260 Broadway, New York—Ramie.

Geo. Carter, 246 Broadway, New York—Linen yarns and thread.

Boston Yarn Co., 50 State St., Boston, Mass.—Cotton yarn.

Wonalancit Co., Nashua, N. H.—Samples of cotton.

Botany Worsted Mills, Passaic, N. J.—French spun worsted yarn.

C. E. Riley, 65 Franklin St., Boston, Mass.—Yarns and fabrics.

Adirondack Wool Co., Little Falls, N. Y.—Wools and shoddies.

Sutcliffe, Atlantic Ave., Boston, Mass.—Foreign wools.

Francis Willey & Co., 556 Atlantic Ave., Boston, Mass.—Top, foreign wools.

John L. Farrell, 210 Summer St., Boston, Mass.—Mohair, noils, and carpet wools.

The J. R. Montgomery Co., Windsor Locks, Conn.—Novelty yarns.

Catlin & Co., 67 Chauncey St., Boston, Mass.—Cotton yarns.

Norfolk Woolen Co., Franklin, Mass.—Shoddies.

Parker & Wilder Co., Boston, Mass.—Samples of fabrics.

Lawrence & Co., Franklin St., Boston, Mass.—Samples of fabrics.

Joy, Langdon, & Co., Boston, Mass.—Samples of fabrics.

Clark Thread Co., Newark, N. J.—Exhibit.

George A. Clark & Bro., 400 Broadway, New York—Cabinet and booklet.

[Pg 321] Cheney Bro., So. Manchester, Conn.—Silk samples, silk fabrics.

Johnson & Johnson, New Brunswick, N. J.—Wall chart of cotton field.

Scordill, 902 Canal St., New Orleans, La.—Cotton postal cards.

Storey Cotton Co., The Bourse, Philadelphia, Pa.—Booklet, *All about Cotton.*

White Oak Cotton Mills, Greensboro, N. C.—Stereoscopic views.

Willimantic Thread Co., Willimantic, Conn.

Flax Spinning Co., York St., Belfast, Ireland.—Prints illustrating linen manufacture and samples.

Jas. McCutcheon & Co., 5th Ave. and 34th St., New York.—Flax cabinet.

The Linen Thread Co., 96 Franklin St., New York.—Flax cabinet.

Belding Bro. & Co., 526 Broadway, New York.—School exhibits of silk.

Brainerd & Armstrong, 100 Union St., New London, Conn.—Book and cabinet of silk.

Champlain Silk Mills, Whitehall, N. Y.—Spun silk and exhibits.

M. Heminway & Sons, Silk Co., Watertown, Conn.—Booklets on silk.

Nonatuck Silk Co., Florence, Mass.—Sealed cabinets and books on silk.

William Skinner & Sons, 47 East 17th St., New York City.—Silks and satins.

S. Miller, 304 Second Ave., New York.—Wool fiber.

Milton, Bradley Co., Springfield, Mass.—Sheep chart.

A. H. Post, Quaker Hill, New York.—Raw wool by the pound.

Schermerhorn & Co., 12 West 33d St., New York City, N. Y.—Wall chart illustration of sheep.

L. S. Watson Mfg. Co., Worcester, Mass.—Hand cards.

Howard Bros., Worcester, Mass.—Hand cards.

Prin. Columbus Industrial School, Columbus, Ga.—Samples of cotton plant or bolls.

[Pg 322]

Woolen Yarns

Horstman Co., 5th and Cherry St., Philadelphia, Pa.

Lion Yarn Co., 408 Broadway, New York.

Catalogue of Woolen and Worsted Machinery

C. G. Sargent's Sons, Graniteville, Mass.—Wool machinery.

Davis & Ferber Co., No. Andover, Mass.—Woolen and worsted machinery.

Lowell Machine Shop, Lowell, Mass.—Worsted machinery.

Crompton & Knowles, Worcester, Mass.—Worsted silk machinery.

Speed & Stephenson, 170 Summer St., Boston, Mass.—Textile machinery.

George Gerry & Co., Athol, Mass.—Shoddy machinery.

Tolhoust Machine Works, Troy, N. Y.—Hydro extractor.

Parks & Woolson Machine Co., Springfield, Vt.—Machinery.

Curtis, Marble Co., Worcester, Mass.—Finishing machinery.

General Electric Co., 84 State St., Boston, Mass.—Pictures, showing textile machinery in operation by motors.

Hopkins Machine Works, Budgeton, R. I.—Finishing machinery.

Knitting Machinery

Scott & Williams, 88 Pearl St., Boston, Mass.

Nye & Tudick Co., Philadelphia, Pa.

Chemicals, Dyestuffs, and Sizing Materials

The Arabol Mfg. Co., 100 Williams St., New York.—Sizing materials.

Cassella Color Co., 182 Front St., New York.—Coal tar products, dyestuffs, and literature.

Arnold Hoffman & Co., Providence, R. I.—Sizing materials.

H. A. Metz & Co., 122 Hudson St., New York.—Dyestuffs and literature.

Badische Co., 128 Duane St., New York.—Dyestuffs and literature.

[Pg 323]